WORLD TRAVELS WITH A PERIPATETIC MARINE SCIENTIST

By
Carl J. Sindermann, Ph.D.

Print information available on the last page

Rev. date: 10/20/2015

To order additional copies of this book, contact:
Xlibris
1-888-795-4274
www.Xlibris.com
Orders@Xlibris.com

*A NOTE ABOUT THE TITLE

*I have had advice, mostly negative, about using the word "peripatetic" in the title of this book but, obviously, I have persevered. I have done so principally because it is a correct descriptor of my approach to a very important aspect of the role of scientific research laboratory director: frequent contacts with the rest of the scientific world outside the laboratory walls. The present book can thus be considered as an important adjunct to my 2012 book, "The Scientific Research Laboratory Director" (Xlibris, 2012). Together, the two volumes summarize many of my experiences as a scientific laboratory director, from two completely different perspectives, those within the laboratory walls and those outside those walls in a very wide scientific world!

To conclude this brief tirade: "Peripatetic" as I define it <u>should include</u> (in the case of a laboratory director): "<u>Effort required to remain current with research results throughout the world in a specific area of specialization.</u>"

FRONTISPIECE:

Carl J. Sindermann is a former director of the Middle Atlantic Coastal Fisheries Research Center of the National Marine Fisheries Service (NMFS), a division of the National Oceanic and Atmospheric Administration (NOAA). He has had earlier professional experiences as Director of the Tropical Atlantic Biological Laboratory of NMFS in Miami, Florida, and as Director of the Biological Laboratory of NMFS in Oxford, Maryland, following a research career in Boothbay Harbor, Maine, after receiving his Ph.D. from Harvard University.

He has published more than 50 technical papers and 15 books about science and scientists. This book is a description of some of his travels – official and unofficial – to various parts of the world during that long and pleasurable career in ocean science.

THE UNLIKELY BUT TRUE STORY OF THE VERY ABNORMAL METAL FISH PICTURED ON THE COVER

Regardless of the main story line of this book, if any exists, the story of my tumorous and ulcerated metal fish (the cover photograph) must be told in these opening pages. It is a genuine work of art, produced by a skilled artisan in a small shop in downtown Charlotte Amalie in the Virgin Islands. It is a production that I immediately recognized for what the artist was trying to visualize: "the effects of disease in marine fish," which had been a large part of my scientific existence for all the decades of my professional life! I had to buy it at any price, and I did, although with subsequent family migrations it was lost for a long time, and only recently recovered by my son Carl in time to be refurbished by my son Dana and then photographed for the cover of this book.

The fish belongs in this book, as a fitting, indeed a remarkable symbol of the confluence of a career in fish pathology with a career-long exploration of the known world! It is an unlikely but very compatible association with which I am well-pleased and forever grateful. It was focused, to a large extent, on the somewhat obscure scientific specialty of fish pathology that opened many doors for me throughout the world, as described in this book.

CONTENTS

PHASE I: THE BOOTHBAY HARBOR, MAINE FISHERIES RESEARCH LABORATORY, 1956-1959. Early fish disease research

PHASE II: THE OXFORD, MARYLAND BIOLOGICAL LABORATORY, 1960-1964. Shellfish research in the Eastern United States and travel to the Far East and Australia

PHASE III: THE MIAMI, FLORIDA TROPICAL ATLANTIC BIOLOGICAL LABORATORY (TABL), 1965-1970. High seas tuna research off the west coast of Africa and tropical aquaculture in the United States

PHASE IV: THE SANDY HOOK, NEW JERSEY MIDDLE ATLANTIC COASTAL FISHERIES RESEARCH CENTER, 1971-1990. Fish disease and marine pollution studies collaboration with European scientists

PHASE V: TECHNICAL AND NON-TECHNICAL WRITING, 1991 – Present, And General Conclusions About Scientific Travel

INDEX TO FIGURES IN THIS BOOK

INDEX TO MAPS IN THIS BOOK

PROLOGUE:

WHY WRITE A SCIENTIFIC TRAVEL BOOK?
OR, HOW THIS STORY CAME TO BE WRITTEN

Travel books written by authors with many kinds of backgrounds are abundant, to say the least, except for travel books written by scientists who attend international meetings. These tend to be very scarce.

This book follows the predictable path of a relatively unworldly scientist, sometimes accompanied by his wife, on a panoply of excursions over much of the world, mostly with a scientific motivation and justification. The theme, if any, that should emerge from these pages, is that, <u>whereas the scientific profession is spectacular and satisfying in itself, it also provides entry into an international community that is equally exceptional and should be made known to everyone, especially other scientists.</u>

Our occasional joint travels (my wife, Joan, accompanied me on some trips), as described briefly in this book, began erratically when our children were small (we were fortunate to have had excellent and reliable in-home care service always available), and this continued during my early career as a research biologist and laboratory director. Our kids, now grown of course, don't seem to have suffered greatly from our infrequent and not usually extended disappearances to other parts of the planet, where we met and enjoyed the company of innumerable fellow scientists, especially those who were members of scientific working groups or advisory committees that met annually in different countries. That, the foreign personal contact aspect of participating with scientists from many other countries in the affairs of science, has added an entire new dimension to our lives and is an element of foreign travel that is generally underemphasized. Every developed country in this world has a complement of good scientists in every discipline who should be approached, brought into group discussions, and befriended! I have found that the best approach to this action is through participation in technical advisory groups that report to international organizations of all kinds, governmental or not. This venue is where much of the international scientific interaction actually takes place, and much more will be said of it later in the book.

Understand clearly that in my focus on the significance of scientific advisory groups in promoting personal interactions between U.S. and foreign scientists, I am by no means reducing the value of international scientific societies like the World Aquaculture Society in developing a truly international membership and holding superb annual meetings in many countries. The combination of international scientific advisory groups and effective international societies can provide an interaction that is optimal. The technical societies provide the base for international communication among scientists, whereas the technical advisory function assures that the contributions of science to multi-national issues are presented and heard. I have participated in both kinds of international scientific activities, and I find them both essential to the worldwide progress of science!

PREFACE: ON THE STRUCTURE OF THIS BOOK

I have puzzled for some time now about the best method of presenting my foreign travel experiences within a framework that would be understandable and even interesting to the reader. The eventual answer that I found was <u>to use as a framework the sequences of my tenures at various laboratories on the Atlantic Coast of the United States</u>, beginning with Phase I, the Fisheries Research Laboratory at Boothbay Harbor, Maine, and then moving on to Phase II, the Biological Laboratory at Oxford, Maryland, and then to Phase III, the Tropical Atlantic Biological Laboratory at Miami, Florida and then to Phase IV, the Middle Atlantic Coastal Fisheries Research Center at Sandy Hook, New Jersey.* Each move required a change in research perspectives because of ongoing and expected program emphasis, and each move resulted in the development of new contacts and responsibilities.

<u>So that is how I have structured the book... on a timeline dictated by successive tenures at each laboratory</u>, followed by an add-on period (Phase V) for contemplation and writing, with an Intergovernmental Personnel Act appointment. Fortunately for this proposed framework, the five career phases, the foreign travel expectations (with some exceptions) were distributed geographically according to needs of the research programs at each location.

Accepting this basic structure for presenting a travel book, and recognizing the many "outliers" that may exist, this is a rough outline of travel presentations that follow in this book:

Phase I 1956-1959	Early years at the Boothbay Harbor, Maine Fisheries Research Laboratory with research on marine fish populations and diseases.
Phase II 1960-1964	Shellfish and marine aquaculture research at the Oxford, Maryland Biological Laboratory.

Phase III 1965-1970	High seas tuna research and marine aquaculture research at the Tropical Atlantic Biological Laboratory in Miami, Florida.
Phase IV 1971-1990	Ocean pollution and fish disease research at the Middle Atlantic Coastal Fisheries Research Center at Sandy Hook, New Jersey.
Phase V 1991-present	Research and writing at Sandy Hook, NJ, Miami, FL, and Oxford, MD with contributions from Intergovernmental Personnel Act appointment. **

Despite this apparent structure, there will be outliers in the following text, such as a proper location to discuss trips to Australia, Iceland, and Central America, but the main flow of the narrative will be as indicated here, from early experiences in Canada and Europe (Phase I) on to several visits to southeastern Asian countries (Japan, Taiwan and South Korea [Phase II], then to Africa (Phase III), then to concentration on European and Japanese interactions (Phase IV), with a final phase (V) reserved for contemplation and writing at several locations.

--

*   Maps of principal travel, charting each phase, are indicated here to further explain the coordination of career phases and principal foreign travel, and individual maps are also reproduced at the beginning of each phase in the text presentation.*

--

--

*   ** The provisions of the Intergovernmental Personnel Act permitted me the utmost freedom to add to the scientific literature in my areas of technical specialization, which I did, in areas of marine fish and shellfish diseases and marine pollution. See reference section.*

ABOUT THE AUTHOR

Our principal traveler and author of this book, Carl J. Sindermann, spent his boyhood in the very small Western Massachusetts town of Blackinton, too small then and now for inclusion on any official state road map. After service as an Army infantry platoon medic in Europe during World War II, and undergraduate studies at the University of Massachusetts, he completed graduate work for the PhD at Harvard, where he fell into the hands of exceptional teachers and outstanding research scientists. He emerged with the PhD, but also an abiding interest in the biology of marine fish and shellfish, an interest he retained throughout his long career as a scientist and administrator.

Dr. Sindermann, early in his scientific career, sampled and enjoyed a few years of full-time university teaching and research on the faculty of Brandeis University after receiving his PhD from Harvard in 1953, but then decided to join the marine research component of the Federal Bureau of Commercial Fisheries, where he enjoyed a long and very productive research and administrative career of 35 years in several east coast fisheries research laboratories, most of them as laboratory director or research center director. During this period he published more than 100 technical papers and reviews, as well as 20 books on technical subjects or non-technical subjects related to science.

His major technical publications concerned diseases of marine fish and shellfish, including his major contribution to science: a thousand-page, two-volume treatise on "Principal Diseases of Marine Fish and Shellfish," and several other volumes on "Ocean Pollution, Coastal Pollution, and Diseases in Marine Aquaculture."

He retired in 1990 with much left to say but still unsaid. He is catching up quickly in retirement with the 2001 revision of his earlier book *Winning the Games Scientists Play* (Perseus) and with the publication of *The Scientific Research*

Laboratory Director (XLibris) published in 2012; and with the current book: *World Travels with a Peripatetic Marine Scientist.*

This is the seventh book in a series about scientists published by Carl J. Sindermann, Ph.D. (some with co-authors). He is a marine biologist who has also written technical books as well as more than 100 technical papers and reviews in his technical specialty.

Other books in this series include:

1. Winning the Games Scientists Play (Plenum, 1982), Revised, Perseus 2001,

2. The Joy of Science (Plenum, 1985),

3. Survival Strategies for New Scientists (Plenum, 1987),

4. The Woman Scientist (with Clarice C. Yentsch, Plenum, 1992),

5. The Scientist as Consultant (with T. K. Sawyer, Plenum Trade, 1997),

6. The Scientific Research Laboratory Director (Xlibris, 2012).

Note: All the books in the series are available through the internet, or from the library of the Rosenstiel School of Marine and Atmospheric Sciences, University of Miami, Rickenbacker Causeway, Miami, Florida.

* A PERSONAL NOTE FROM THE AUTHOR ABOUT HIS FAMILY HISTORY AS A BACKGROUND FOR FOREIGN TRAVEL

My wife Joan and I were born a week apart in the small city of North Adams, in the Berkshire Hills of Western Massachusetts. We went to high school together and married in 1943, during World War II, in which we both participated actively – she as a Navy WAVE with stateside hospital duty and I as an infantry platoon medic in Europe. After the war I enrolled at the University of Massachusetts in Amherst, and our two daughters Nancy and Jeanne were born in nearby Northampton during my undergraduate days. After receiving my BS in Zoology we moved to Cambridge and Harvard, where I was awarded a teaching fellowship at the College and also at the Medical School. I also began summer research in marine biology at the Boothbay Harbor Biological Laboratory in Maine. During that period our first son James was born in Cambridge and our second son Dana in Maine during one of our "research summers" there.

I received my Ph.D. in Biology from Harvard in 1953 and was awarded a faculty teaching position at Brandeis University in Waltham, Massachusetts, teaching invertebrate zoology and general education biology. Joan enrolled at Boston University.

After several pleasant and productive years at Brandeis University I decided to leave academia for a full-time research position with the federal government at the Boothbay Harbor Laboratory in 1956, (a decision that I have never regretted). Our third son Carl was born after we had moved to the Maine coast on a full-time basis, completing our marvelous and planned complement of five intelligent children for which we have been forever grateful. All of them went on to college and to careers of various kinds thereafter.

I was appointed Director of the federal Oxford (Maryland) Biological Laboratory in 1959, and served there until 1965 when I was appointed Director of the Tropical Atlantic Biological Laboratory in Miami. Then in 1970 I was appointed Director of the

Middle Atlantic Coastal Fisheries Research Center at Sandy Hook, New Jersey, where I remained until retirement in 1990. All these moves involved progressively broader research, administrative responsibilities and travel – all of which I enjoyed immensely.

So there we were, me with a Harvard degree and a series of long-term federal research/administrative positions along the Atlantic Coast, us with five super-bright children. What was the totality of our world to be like? Part of it – but only the briefest external part of it – foreign travel – is sketched in brief episodes on the pages that follow. There is so much more that is not included.

In our own lifetime journey together, Joan and I have left parts of our family's past and parts of our own lives in small towns like Boothbay Harbor, Maine, Oxford, Maryland and Amherst, Massachusetts, and much smaller parts in a much wider world of foreign meetings and foreign travel – all in what we considered a reasonable balance. We, Joan and I, are happy with that balance!

A FAMILY TRIBUTE

This book includes brief encounters with parts of the world where my wife Joan and I were able to travel together. She was always an excellent traveling companion – ready for whatever the new day had to offer. Needless to say, those trips we made together were the best trips of all that I made.

We did, however, leave behind at home in very good hands, of course, a growing family of girls and boys – not often, but occasionally. They have come together as adults to help me in many ways with the completion of this book.

Our two daughters, Nancy Sweet and Jeanne Kennedy have acted as reviewers of the manuscript, thereby improving it immensely. Our son Jim prepared all the maps for the book, and our sons Dana and Carl helped with the logistics of publishing a book of this kind and interacting with the publisher.

So it was a family production, and I am happy with it. Writing the book has of course brought to mind many pleasant places and days and people, so the whole exercise has been another pleasant journey for us, and we hope readers will find pleasure with it too.

INTRODUCTION
THE GENESIS OF A SCIENTIFIC TRAVEL BOOK

I have been traveling to foreign countries for almost half a century, sometimes with my wife Joan, mostly to carry out functions that I considered important to my position of scientific research laboratory director at a number of federal research laboratories located along the Atlantic Coast of the United States. Such travel was, of course, only a minor part of my total responsibilities as a scientist and a research laboratory director. My most important responsibilities as director were to develop and maintain a productive research organization, and to maintain personal competence and visibility simultaneously as a scientist.

One of my original hopes for this book was that it would explain clearly, by example and opinion, what the possible role for an American scientific research laboratory director should be and can be in research areas that have obvious international counterparts, and can benefit from foreign interconnections in research.

Another hope, encouraged by my association with international statutory organizations in an advisory capacity [International Council for the Exploration of the Seas (ICES] and United States – Japan Joint Panels on National Resources (UJNR)], is that the international role for a U.S. research laboratory director is clearly defined and available.

I say these things with the clear codicil that <u>the primary responsibility of any research laboratory director is development and maintenance of a productive scientific research laboratory at home, which demands the exceptional organizational and communicative capabilities of the director. Once these favorable conditions prevail, and only if they prevail, can the director then depend on the presence and functioning of an effective Assistant Laboratory Director, who is fully capable of maintaining research productivity and morale at high</u>

<u>levels in the director's absence</u>. If that kind of person is not available, then any travel or absence of the director from the laboratory should be limited to very short periods.

This book of travels that I have enjoyed exists only because I have had the great good fortune of having had several excellent Assistant Laboratory Directors, at several laboratories, who permitted me the freedom of occasional foreign travel, a larger responsibility for them than might be anticipated, since this was the only condition under which all the wonderful foreign adventures reported in this book were in any way possible for me.

In addition to these basic responsibilities were, of course, lesser ones, still of importance, that required reasonable amounts of travel, both domestic and foreign. Since this book concerns <u>foreign</u> travel, I should list the important reasons, scientific and administrative, for my presence, over time, in so many interesting parts of the planet other than our own great country. Here are the principal ones, from my perspective as a retired research laboratory director and world traveler:

- To present personal research and review papers at international scientific symposia;

- To present results of laboratory research on large multi-program studies;

- To attend planning meetings with foreign laboratory directors who are cooperating with American laboratory directors in joint research programs;

- To present technical information at statutory meetings*, in which the United States is a participant;

- To act for ten years as Chairman of a Working Group on Introduced Species of the International Council for the Exploration of the Sea (ICES);

- To serve for twelve years as panel member (six as editor of proceedings) of the United States – Japan Joint Panels on Natural Resources (UJNR);

- To present scientific papers and review papers at meetings of international professional societies (International Society of Parasitologists, World Aquaculture Society, as examples);

- To attend foreign scientific meetings at requests from foreign national academies of sciences;

and;

- To visit foreign sites for production of living organisms from coastal waters (such as oysters or mussels) that have been proposed to be introduced into the United States.

--

* The term "statutory", as used in this book, refers to research conducted in response to needs or requirements of international treaties or other agreements.

--

This book, as you will discover, describes an example of the role expected of today's scientific research laboratory director insofar as international travel is concerned. It assumes a stable productive home laboratory, an excellent deputy, and the existence of traditional methods of inter-laboratory communications, including electronic ones, but points to the <u>personal</u> methods of direct communication on an international basis, for which there is no substitute: the <u>personal</u> participation in international professional society meetings; the <u>personal</u> participation in technical working groups and advisory groups of international government statutory organizations; and the <u>personal</u> active roles in international science that matters (such as global warming or species extinctions or habitat degradation).

Before I conclude these introductory comments about foreign travel as a scientific research laboratory director, I want to emphasize a few points that should become apparent as the chapters roll on:

- that a reasonable amount of travel, foreign and domestic, is an expected and necessary part of the position of scientific research laboratory director;

- that such travel should have purposes that are of scientific and/or administrative nature, or both;

- that travel commitments of a scientific research laboratory director should never become so intense as to affect the research productivity of the home laboratory negatively; and

- that, conversely to the preceding item, the research laboratory and its director should never be allowed to become isolated from the vigorous onrush of science, foreign or domestic, by any failure to participate in professional activities beyond the institution's walls;

- and, finally, that travel – particularly foreign travel – one of the most favorable elements of a scientific research laboratory director's job -- is to be accomplished with as much skill and grace as possible! The competition out there beyond our national boundaries can be severe.

So, with these preliminary words of advice, opinion, and caution, let's be off for a brief summarization of some foreign adventures of one scientific research laboratory director, and often his wife Joan, in a surprising number of interesting places on this planet.*

During my entire marine research career I was employed by the United States Federal Government in a series of line agencies variously named Bureau of Commercial Fisheries, (BCF) or National Marine Fisheries (NMFS) Service or National Oceanic and Atmospheric Administration (NOAA). Regardless of their names, ocean fisheries research was always an important element of the annual budget – one that supported the programs of coastal fisheries research such as those that I directed with such pleasure along the Atlantic Coast of the United States.

Part of my perceived responsibilities as a federal research laboratory director was to find and maintain productive contacts with counterpart research groups in other developed and developing countries – a role that I enjoyed and considered relevant to my research position in the United States. This book contains many of my findings and observations during foreign travel – beyond the usual tourist impressions (which are included too).

--

* I have tried to organize this chronicle of foreign travel in <u>phases</u> (with maps) that are related to the Atlantic Coast laboratories at which I was based and for their research missions at that time. This makes sense to me, but may be <u>very</u> confusing to the casual reader. If it all seems too complex, just remember that a research and administrative objective existed for each visit, and that each one was a learning experience for me, so enjoy!

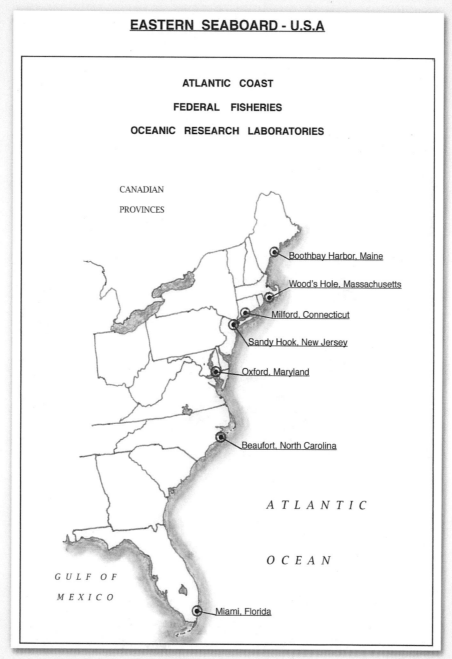

EASTERN SEABOARD - U.S.A

ATLANTIC COAST

FEDERAL FISHERIES

OCEANIC RESEARCH LABORATORIES

CANADIAN

PROVINCES

Boothbay Harbor, Maine

Wood's Hole, Massachusetts

Milford, Connecticut

Sandy Hook, New Jersey

Oxford, Maryland

Beaufort, North Carolina

ATLANTIC

OCEAN

GULF OF

MEXICO

Miami, Florida

Map 1. Federal Marine Biological Research Laboratories on the U.S. Atlantic Coast

PHASE I
THE BOOTHBAY HARBOR, MAINE
FISHERIES RESEARCH LABORATORY
(1956-1959)

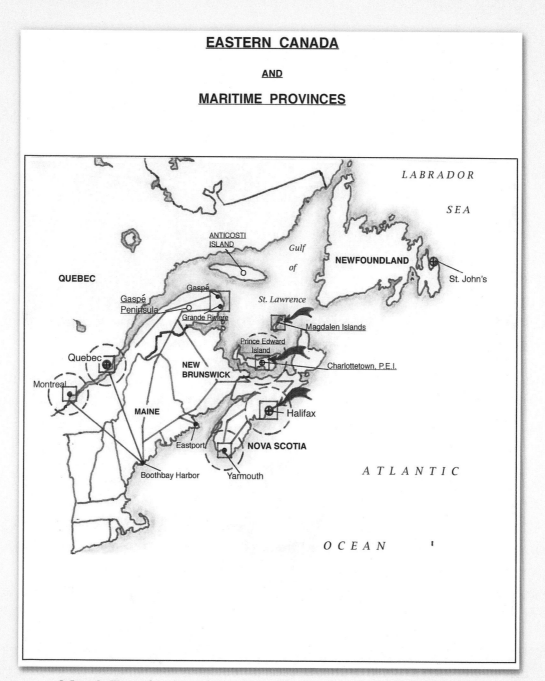

Map 2. Travel in Northern New England and Eastern Canada.

PHASE I
EARLY FISH DISEASE RESEARCH AT THE BOOTHBAY HARBOR, MAINE FISHERIES RESEARCH LABORATORY
1956 - 1959

After acquiring a Ph.D. in Biology at Harvard in 1953 I spent several years as an Instructor in Biology at what was then a very new and rapidly developing university, Brandeis University in Waltham, Massachusetts. There I taught courses in General Education Biology and Invertebrate Zoology. I also continued marine research on fish diseases and schistosome dermatitis during summer vacations in Maine, at the Boothbay Harbor Fisheries Research Laboratory.

I then decided (in 1956) to leave Brandeis and to join the research staff of the Boothbay Harbor Fisheries Research Laboratory (a federal government laboratory) as a research biologist and Program Leader in Fish Pathology. My very productive tenure in that position was for four years, from 1956-1959.

Of all my research assignments on the Eastern seacoast with the federal government, the fisheries research laboratory at Boothbay Harbor on the Maine coast was, for its purpose, the best of many good ones. The laboratory was in a historic building, located on a scenic harbor in a scenic village in mid-coast Maine. Our house, sufficiently spacious for a growing family, was on a hill next to the town's storm warning tower and overlooking the harbor. Best of all, I was doing research directly related to my earlier Harvard Ph.D. studies (fish pathology).

As the future emerged, each phase of my professional career also required relocation of my family and transition to a new community. So in 1956, my wife Joan, three small children and a dog made a definitive trip in the family station wagon from Cambridge to the small harbor town of Boothbay Harbor, leaving behind all the city's academic and other resources, for life in a small coastal community. Joan was my greatest supporter in making that and all the later transitions that accompanied the development of my professional career.

My early full-time research experiences at the Boothbay Harbor Laboratory were materially enhanced by the presence of the Director and Chief of Investigations, Mr. Leslie W. Scattergood, an excellent scientist and a wise supervisor. As a program leader from 1956-1959, I had the utmost freedom in developing and executing research programs in relevant fisheries areas, including joint international projects with Canadian research biologists employed by provincial governments and the Fisheries Research Board of Canada. It was in this early period that my joint international studies in fish and shellfish pathology began – studies that led me later to many places in the world.

Figure 1. Our young family gathers at our hilltop house in Boothbay Harbor, Maine.

CHAPTER 1
EARLY RESEARCH IN NORTHEASTERN NORTH AMERICA: MASS MORTALITIES OF FISH IN THE GULF OF ST. LAWRENCE

My first professional encounters with so-called "foreign scientists" occurred reasonably close to the United States in a country that was largely English speaking (except for the area of my research), Canada. Early in my career as a research biologist with the federal government in New England I encountered, as did Canadian scientists, widespread problems with marine fish mortalities due to epizootic disease. The mortalities were most severe in the Gulf of St. Lawrence fish populations, to the point where catches were severely reduced for several years in the mid-1950s, so a joint international scientific study seemed to be in order and was eventually endorsed by the United Nations (FAO) International Commission for the Northwest Atlantic Fisheries (ICNAF). Research groups from the two most affected member countries (Canada and the United States) committed research staff members to the project for the several years of the epizootic period, and much was learned and published about the true role of disease in the sea. It was for me a time of the closest cooperation between scientific groups from the two nations that I have ever witnessed, and an excellent example of what is feasible internationally. Our fish sampling trips in Canada included the Atlantic coast of Nova Scotia in late autumn, where we watched the keel laying and early construction in the town of Shelburne, of the replica of the famous pirate sailing ship "Bounty." This ship was later sunk off the coast of North Carolina during the "Superstorm Sandy" in 2012 (and not burned off Pitcairn Island as depicted in the earlier movie starring Marlon Brando, its movie "mutineer captain" who was strongly opposed to that fate for "his" ship).

Much of the real action took place in the Gulf of St. Lawrence, where the disease had reached its epizootic peak, and where field work was most intensive. This meant long, long road trips from Maine through New Brunswick to Quebec's Gaspé Peninsula and beyond for U.S. participants, but there was a genuine sense of involvement in something really large, scientifically as well as geographically. That made the whole international venture worthwhile to all of us.

The road trips that I just described, inadequately, as "long" were really much more than that. They took us from Maine through the evergreen forests and then the north coast of the Province of New Brunswick, Canada, then along the stunning southern coast of the Gaspe Peninsula of Quebec, to our temporary field station at the Grande Rivière Provincial Fisheries Research Laboratory. We (I and a technician or with my wife Joan) made these trips in the spring and sometimes autumn for several years during the Gulf of St. Lawrence epizootic period, and we thoroughly enjoyed every one, as scientists and as international travelers in an exceptionally interesting and scenic part of the world.

Yes, the trips were long and tedious, but they took us through some of the most spectacular countryside in Eastern North America – the shoreline of the Canadian Maritime Provinces including the Gaspé Peninsula, where we became briefly, part of the community of scientists in a remote provincial laboratory during our field work.

Our joint research activities in Canada, particularly those at the Grande Rivière Provincial Laboratory on the Gaspé Peninsula in Quebec, where French was normally spoken, and only the Director spoke English, were instructive for all of us involved. It was especially important for me to realize the international nature of the work we were doing and the potential significance of factors such as diseases, from country to country, regardless of the language spoken. In subsequent years and in different interactions with foreign scientists I did observe, though, that many scientists were familiar to varying degrees with the

English language and English technical literature. Despite language differences, we were fortunate to be able to establish, then and later, close professional and even personal relationships with many extraordinary scientists around the world.

We were able to produce a number of scientific papers based on the disease studies in the Gulf of St. Lawrence in national and international journals, and the information about the disease proved useful in understanding a similar epizootic event that occurred in sea herring of the European coasts decades later.

We (I and another biologist or technician, or on some occasions my wife, Joan) took this long, long trip by car from Boothbay Harbor on the Maine coast to the Gaspé Peninsula in spring and autumn to coincide with the annual fish migrations of herring. Working closely with our Canadian counterparts from the St. Andrews (New Brunswick) Laboratory of the Fisheries Research Board of Canada, we were documenting the course and population impacts of a lethal fungal disease on sea herring and related species. From a research perspective, this was a truly exceptional opportunity for joint international research leading to better understanding of the role of epizootic disease in the sea. For me, the experience offered an early and important introduction to international science and to some of the organizations that permit its functioning (The U.N. [FAO] International Commission for the Northwest Atlantic Fisheries [ICNAF] and the International Council for the Exploration of the Sea [ICES]).

Those early research adventures, in the northern reaches of North America, seemed to set a pattern of cooperative studies with foreign scientists with me. Or course it was relatively easy in this case since most of our contacts were English or French speaking or with those who understood English quite well, and Joan both understood and spoke French (still a very important language skill in some of the fishing villages on the Gaspé Peninsula and the Magdalen Islands in the central Gulf of St. Lawrence).

I can't withdraw from this discussion of the Gulf of St. Lawrence without brief mention of a scientific laboratory which we visited repeatedly that made a significant contribution to humanity over a half century ago. This is a tiny Provincial Fisheries Laboratory on one of the very isolated Magdalen Islands group in the central part of the Gulf of St. Lawrence. This laboratory is probably one of the most isolated of any research laboratory that I have ever visited in my entire career. It is accessible by boat when ice does not cover the Gulf, or by small plane when the hard sand beach is clear. The early research programs of this laboratory concerned lobsters, herring, and harp seals. Its principal contribution to science has been to harp seal biology. Its principal contribution to <u>humanity</u> has been provision of data that resulted in the <u>prohibition</u> in 1982 by the Canadian Government of the killing of harp seals on the ice of the Gulf of St. Lawrence for their fur. This very tardy prohibition resulted from the totally inhumane practices of hunters who <u>killed pregnant females</u> on the ice and then killed and stripped their unborn young of their pure white fur. An almost unbelievably cruel practice! Now, belatedly, outlawed!!

I must, of course, point out that all this work with the Canadian herring fungal epizootic was only a small, if important, part of my research functions at the Boothbay Harbor Fisheries Research Laboratory. Under the far-sighted directorship of L.W. Scattergood I was given truly remarkable freedom of choice in selecting and pursuing research programs in marine fish population and disease problems, with some emphasis, logically, on United States waters of the Gulf of Maine.

It should also be stated here that the lethal fungal disease of clupeoid fishes caused by <u>Ichthyophonus hoferi</u>, with effects described by us and Canadian biologists in the Gulf of Saint Lawrence, is now a world-wide problem. It has occurred at epizootic levels in European herring of the North Sea and Northwest Atlantic, in Alaskan populations of Pacific herring, and even in imported stocks of sea herring used as food for experimental tuna mariculture in Australia!

Figure 2. Carl and Joan, with a friend from Woods Hole, about to depart for our first ICES trip to Europe in 1958.

It was during the first and most productive phase of my research at the Boothbay Harbor (Maine) Fisheries Research Laboratory that first contacts were made, almost accidentally, with the prestigious International Council for the Exploration of the Sea (ICES). Based in Copenhagen and with a complex structure of committees and working groups that provided scientific advice to member countries (the United States was and is one), ICES has a long history of annual meetings, usually on ocean-related subjects important at the moment.

One topic for the year that I first became involved with this venerable scientific organization (in addition to the fish disease problem) concerned methods of distinguishing high seas populations and subpopulations of fish. I had been developing parasitological and immunological techniques relevant to that subject, and our Washington office endorsed my participation to present a paper on my research at the annual meeting to be held that year in Copenhagen. It would be the first of several trips to that delightful city that Joan and I would make and we enjoyed every minute, travelling on one of the early Boeing 747 flights to Europe with another biology couple from the Woods Hole, Massachusetts Laboratory. The meeting was one week long, with suitable ocean-oriented evening entertainment, including a Fisherman's Fair on the opening night, at which we won the grand prize, a 25 pound halibut freshly-caught and wrapped in newspaper. We tried to give the fish away on the street, but were met with blank or disapproving stares, even from kids. We eventually put the damn fish out on the window sill of the hotel (it was October) and eventually solved the problem later in the week by changing hotels and leaving the "damn fish" behind.

Getting back to the meeting, I gave my paper, which as I recall, was titled "Blood types in Atlantic redfish". It was meant as a method of identifying populations and movements of a deep sea fish and it was met with dead silence (probably because it was such a departure from standard fishery population research methodology being employed at the time).

That early invitation helped to foster a career-long association with ICES, which was and is a preeminent long-term presence in marine fisheries research. Joan and I traveled to ICES Working Group meetings held in many European countries, and I frequently presented reports at Council meetings, often in Copenhagen, but also in many other European countries.

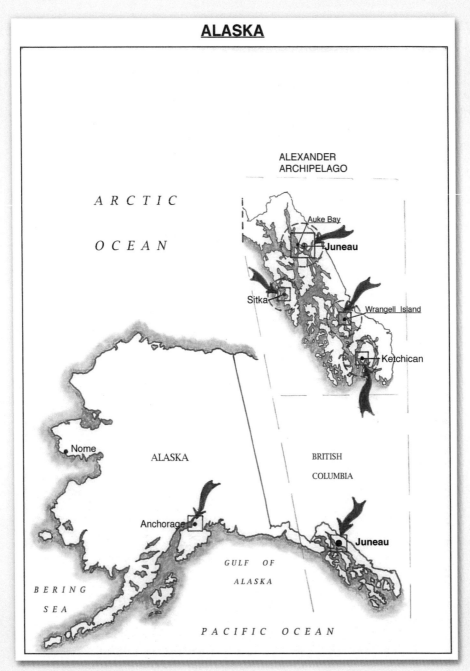

Map 3. Travels in Southeastern Alaska.

CHAPTER 2
A "DREAM TRIP" ALONG THE SOUTHEASTERN COAST OF ALASKA*

In the early 1960s, our central office for fisheries in Washington D.C., in a moment of great wisdom, had decided to form a <u>select committee of research laboratory directors</u> from around the nation's coastline to examine the nature and effectiveness of its National Marine Fisheries Service (NMFS) research programs concerned with salmon and other species in Southeastern Alaska. I was appointed a member of that committee. The programs had their headquarters at the principal NMFS Laboratory at Auke Bay located near Juneau, but, they also operated a number of field stations at Sitka, Wrangell, and Ketchikan. A converted Grumman seaplane, yes, a real seaplane, operated on contract with NMFS, was made available to the committee for the full week of field inspections and staff interviews on the Southeastern Alaska coast.

Imagine if you can, a full week of close interaction with six compatible counterpart laboratory directors travelling in a real seaplane to different designated field stations of NMFS. This is the stuff dreams are made of, at least for scientific administrators like me!

Our method of travel was somewhat unique and perfectly suited for the dreadfully rough coastline of Southeastern Alaska. There we were, six of us from all the coastlines of continental United States, traveling in a converted World War II PBY Grumman seaplane to hard-to-reach research outposts on the Southeastern Alaska coast. We didn't fully appreciate it then, but it was for all of us, or at least some of us, a scientific and administrative adventure of a lifetime. Its basic purpose was to conduct an intensive on-the-scene examination of the entire national research program for Southeastern Alaska's fisheries and to make recommendations for the future of that program. Most importantly, we felt that our views would be considered seriously by decision makers in the organization.

The procedure at each field station was usually the same. We (the Committee) descended on the forewarned and fully prepared field stations in our mighty seaplane to the river or bay adjacent to each remote research facility, where we were met by the chief biologist and his or her staff members for presentations followed by discussions and tours of local equipment and facilities. After final closed discussions within the committee a closeout session was usually held with the biologist in charge and his or her professional staff members. The committee usually then adjourned to a nearby town for rest and relaxation and was ready in the morning for our next seaplane ride to the next field station, to repeat the entire process. It sounds exhausting, and it was, but it was exhilarating too, for all of us. We all knew, however vaguely, that it was a unique and valuable experience in research administration that none of us would ever forget!

Figure 3. Members of the National Program Review Committee with their Grumman PBY in Southeastern Alaska.

--

* *The more observant reader may have noted that I have already departed from my stated structural plan of five phases of the narrative, based on the research laboratory where I was located at the time of the travel. This Alaskan travel fitted so well with my other exploits in Northern North America that I felt a strong urge to include it here in Phase I, even though I was well into Phase II (the Oxford Laboratory) at the actual time of travel.*

--

To return to my story – travels of the Committee on Evaluation of Southeastern Alaskan Fisheries Research took us from our home laboratories to the Fisheries Research Laboratory at Auke Bay in Southeastern Alaska, and then by seaplane to Sitka, Wrangell and Ketchikan, returning to the Auke Bay Laboratory for summaries and preliminary discussions. After this we elected to meet at the Oxford, Maryland Laboratory to draft a final report because of its proximity to Washington D.C. I was at the time a newly-appointed (1962) laboratory director at Oxford.

Unfortunately, and adding to the "other world" like nature of the whole project, the scheduled date for the Oxford meeting (October 12, 1962) preceded by only a few days the October 15 onset of what has become known in the history books as the "Cuban Missile Crisis", when the Soviet Union's intercontinental nuclear missiles were delivered to Cuba, armed and aimed at the United States, with others aboard ships bound from the Soviet Union to Cuba.

At any rate, and despite all of the international furor, our draft report was prepared and submitted, and subsequently disappeared from view! The confrontation with the Soviet Union was reaching its peak when the Alaskan review committee was completing its work at Oxford under threat of nuclear war, but acting as "good bureaucrats" with documents to prepare. Fortunately, as history discloses, the Soviet dictator Nikita Kruschev backed away from nuclear war under pressure from then U.S. President, John F. Kennedy. What history has not disclosed, at least to me, is the fate of that fisheries report that was completed under such stressful circumstances. But the absolute pleasure of one week of airborne freedom with good company in the acquisition of data for that report will remain with me as long as I live, despite the very real concern that we all felt about travel back to our families (mine was still living in Maine at the time) in the event of nuclear war as a consequence of the Cuban Missile Crisis.

So much for my Alaskan adventure – not to be repeated but never to be forgotten!

This unusual visit to Southeastern Alaska was my longest continuous occupancy of that part of our country, except for brief staff meetings at the Auke Bay Laboratory or occasional overnight stopovers in Anchorage during trips to or from the Far East. Despite the brevity of contacts with Alaskans on their own turf, other intense relationships with scientists who were also Alaskans provided some reasonable perspective on an elite group of people that I have assembled in my own mind – "Individuals from around the earth who enjoy life and prosper in geographic environments that are often less than inviting." I have tried to express my reactions to this special group in a small digression, here, titled "Who is the Alaskan?"*

--

*I believe from this brief observation period (and admitting to a substantial degree of favoritism) that it takes a special kind of human being to live and prosper in Alaska – a person equipped with countless abilities and traits that are absent or poorly represented in any general sample of the human population. To be an "Alaskan" is somewhat comparable to being an "Australian", or a "Scotsman", or a "Newfoundlander", or an "Irishman", or a "Nova Scotian". Current members of these highly selected groups are almost immediately *recognizable by a common set of characteristics or traits, each of which may be present to a greater or lesser degree in any individual.*

Clearly, a combination of genetics, nurture, and environment is at work here, in producing the people that I have mentioned above, of which <u>Alaskan</u> is only one category. During my brief contacts with members of that elite world-wide group to which true "<u>Alaskans</u>" belong, I have gradually assembled a list of characteristics, better described probably as "attitudes towards the world". I have found them to embody some or most of these characteristics:

- *They always carry a noticeable piece of their origin proudly and sometimes publicly, but not offensively;*

- *They are usually respectful to those with other origins, except for the ones who put on airs;*

- *They like gatherings of their own kind, but accept temporary guests who laugh a lot and contribute;*

- *They are proud of their country of origin and state of residence, but they recognize that there are other places on the planet that may be reasonably acceptable as well;*

- *They are not confused by today's mechanical or electronic world, and may in fact be better informed than others;*

- *They are invariably open to discussion about controversial issues, but they can be serious opponents in debates;*

- *They are pleased with themselves for being able to survive and prosper in an often inhospitable environment!*

--

Most of the people living in Alaska today (and most of the people we met there) had come from "away" as the expression goes, but they identify themselves as "Alaskans".

Some, though, were born in Alaska, and except for periods at college in the lower states, had spent their entire lives, by choice, in Alaska. It was in this group that we found most of our "Alaskans", although their numbers were augmented by occasional college graduates who were born in the lower states and were in the process of becoming "Alaskans".

--

PHASE II
THE OXFORD, MARYLAND
BIOLOGICAL LABORATORY
(1960 - 1964)

Figure 4. The Oxford (Maryland) Biological Laboratory.

PHASE II
SHELLFISH DISEASE RESEARCH AT THE OXFORD BIOLOGICAL LABORATORY, OXFORD, MARYLAND AND TRAVEL TO THE FAR EAST: JAPAN, SOUTH KOREA AND TAIWAN, AND AUSTRALIA
1960-1965

I was selected in 1960 to be Director of the Oxford, Maryland Biological Laboratory, a federal laboratory engaged in shellfish research, with emphasis on diseases that were destroying much of the populations of Eastern oysters, *Crassostrea virginica*, in the Middle Atlantic states. The laboratory entered a five year active period of staff recruitment, research, and publication on all diseases of shellfish – molluscan and crustacean.

In chapter one of this book I described some aspects of my biological research with sea herring disease in the Canadian Gulf of Saint Lawrence and in the U.S. Gulf of Maine. The work there and subsequent publications probably contributed to my designation as Director of the Bureau of Fisheries Biological Research Laboratory at Oxford, Maryland, where there was already strong program emphasis on the molluscan and crustacean shellfish diseases that were particularly destructive at that time (early 1960's) in the Middle Atlantic States.

My move in the early 1960's to the Oxford Biological Laboratory brought me in contact with a number of Japanese, South Korean and Taiwanese laboratory directors and biologists who were in the process of reconstituting shellfish research and production as major industries, and were interested in exports to the United States. My concern, of course, with possible importation into the U.S. was possible disease transmission from the Orient to our coastal shellfish populations – a concern that did not prevent the early actuality of such an event happening, through actions of private individuals, despite the warnings of scientists to take proper preventive measures with live imports. That disease event (and others) helped to ensure development and approval, at least in European countries, of a "Code of Practice" to ensure prevention of the introduction of non-native pathogens of marine fish and shellfish, by the International Council for the Exploration of the Seas (ICES) – one of the most active international ocean-oriented organizations in the world today!

For me, the words, "Far East" bring to mind memories of specific periods in the 1960's, 1970's, and the 1980's in Japan, South Korea and Taiwan, during a time of rapid intensification of scientific contacts with those countries following a long period of wars and dislocation, and of hardship and starvation. The energies and hopes of the people had been tested and many looked to the United States and the United Nations for support and assistance. We were part of that large support program for the revitalization of ocean-based industries – and shellfish production was and is a major industry in this category throughout the Far East. Disease control was only one minor aspect of our joint activities, of course, since the countries were at the time also interested in large-scale live export programs with the West.

Our repeated trips, especially to Japan and South Korea, enabled us to observe, directly, the success of some of those shellfish production programs – success for the local economy and for the source of export products. During Phase II and even Phase III of my research career, I emphasized shellfish culture and disease, and interacted with a substantial body of research scientists in Japan, South Korea, and Taiwan during what was a critical period of national expansion for each Far Eastern country. Our efforts, though clearly of United States volition, and narrowly focused, can be seen as part of a general post-war recovery period in the Far East.

My limited Far Eastern ventures constituted, in most instances, positive experiences in very foreign scientific and cultural environments. I treasure them all and I would point out that the shellfish disease research at the Oxford

Laboratory and its potential impact on world affairs seemed to justify – at least in my mind – a significant relocation for our family, away from familiar New England roots to the beginning of a "peripatetic" life, with our children maturing in different places along the U.S. Atlantic Seaboard.

As part of the introduction to my tale of travels to the Far East – to Japan, South Korea, and Taiwan – I must also include the story of the travels of the Pacific or Asian or Japanese oyster (<u>Crassostrea gigas</u>), with its acceptance or rejection in oyster-producing countries of the rest of the world (including its continuing rejection on the East Coast of the United States). The history of this hardy but very edible species has been one of endless discussions and scientific studies but <u>no sustainable introductions on the U.S. East Coast</u>, in spite of highly successful introductions on the U.S. <u>West</u> Coast, and in spite of successful introductions in a number of European countries. The U.S. East Coast oyster production still depends on a disease-ridden surviving population of its native oyster <u>Crassostrea virginica</u>! Here on the East Coast of the United States, attitudes about the introduction of Pacific oysters are strangely conservative, despite some incidental experimentation with <u>C. gigas</u> at a number of research facilities. Despite this experimentation, no oysters are permitted for commercial production in natural waters of the U.S. East Coast except for the native oyster <u>Crassostrea virginica</u>.

National attitudes can be important in decisions about introduced species such as oysters. France, for example, imported in the early 1980's many cargo plane loads of <u>Crassostrea gigas</u> on direct flights from the Far East, and planted them in depleted native oyster growing areas. The successful French oyster industry of today depends almost entirely on the introduced species cultivated in near shore beds.

Herein – with oyster disease problems, real or ephemeral, lies part of the reason for a decade of my foreign travel, particularly to the Far East, in search of scientific information leading to solutions to oyster production problems, real or imaginary, originally caused by disease. Against the larger scale of oyster survival or death, and of correct or incorrect choices of oyster species to cultivate for production, is the successful French experience of taking disease risks associated with mass introductions of foreign oysters to create essentially a new industry with a different species, using different production methods (also imported). This explanation may seem somewhat technical for a book about travel, but the story of the current almost world-wide distribution of the Pacific oyster, <u>Crassostrea gigas</u> and the world-wide scientific controversy it generated is too fascinating to be ignored, and too intricately interwoven with my foreign travel to be treated lightly.

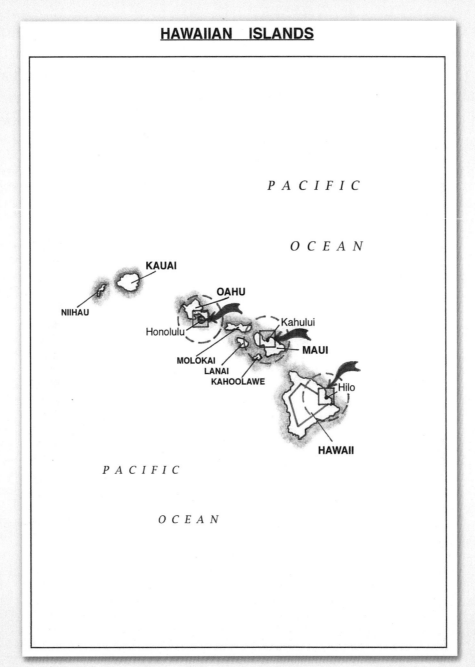

Map 4. The Hawaiian Islands.

CHAPTER 3
HAWAII: AN ISLAND STOPOVER ON THE WAY TO THE FAR EAST THAT BECAME A DESTINATION OF ITS OWN

The Pacific Ocean is obviously huge on anyone's map, but it would seem much larger and less friendly were it not for a delightful cluster of islands, the Hawaiian Islands, located significantly east of its center. Those islands have served as important stopping points for long distance commercial sailing ships of earlier centuries, and now act as centers for trans-Pacific air travel and ship-borne cruise travel. With the acquisition of United States statehood, a welcoming population, and a stable economy, the Hawaiian Islands have become a world choice as a significant destination for high-end leisure travel. It is also a focus for a wide variety of scientific studies: of vulcanism, of high seas pollution, of effects of introduced animal and plant species, of fluctuations in oceanic currents, global warming, migrations of tuna, and many others, conducted by Hawaiian laboratories or by branches of mainland laboratories of Pacific-rim countries.

Hawaii is admirably placed in the Pacific Ocean to be a hub for much trans-Pacific air travel, and has the added value of being an excellent year-round tourist destination for long-distance travelers as well as a perfect leisure stopover for through passengers going to the Far East or coming from the Far East to North America. For me, and selected other travelers in either direction, Hawaii also has the large advantage of being the site of one of the National Oceanic and Atmospheric Administration's (NOAA) major laboratories, with broad ocean resource programs all located near Honolulu on the island of Oahu, with a standing welcome to NOAA mainland staff members passing through.

The island of Oahu is itself a fascinating place, beginning with the Waikiki Beach area with overwhelming shorefront hotels, the beach itself, and the many pleasures and treasures to be found on Khalikaua Avenue which parallels the beach. Then beyond the beach is an entire island of beautiful countryside, with other beaches that are lightly populated with scene after scene of tropical splendor, interspersed with small towns and memorials of World War II, a war that began on this island so many years ago.

In my opinion, through passengers should never forego an opportunity for a stopover of at least two days in this delightful place, even if it is at a large extra cost. It is truly one of the world's treasures.

Hawaii (or rather the Hawaiian islands) begins, of course, with an obligatory several days in an oceanfront hotel on Waikiki Beach on the island of Oahu, and ends much later with departure from the spectacular "big island" of Hawaii, with several other island choices inserted between.

We have visited Hawaii mostly on brief official visits or occasionally for week-long scientific meetings (our favorites). Such visits are always pleasant and always too short. We leave with vows to return, and to stay as long as possible!

Hawaii is noted as a desirable location for the annual meetings of many scientific societies, including the World Mariculture Society, which is in turn noted for selecting exotic locations for its annual meetings and carrying them off with great splendor (and with much good science too). The Society met there in 1980. *

What I remember best about that meeting (other than the good science imparted) were the evening hospitality sessions in dramatic locations in around Honolulu, including one impromptu evening reception aboard a Belgian Navy destroyer that just happened to be in port (don't ask me why) and whose captain welcomed us aboard as a gesture of international cooperation. A truly great evening was enjoyed by an international group gathered very tightly (physically) for cocktails and "aquaculture discussions" held in the destroyer's tiny main wardroom – with no-one ready to go ashore readily and more and more participants arriving – even some Russian sailors from an adjacent ship, ready to party. It was a night to remember and a meeting to remember! We (Joan and I) met some of those same Russian sailors on Waikiki Beach the next day and they seemed to be appalled to see Joan in a bathing suit (she was somewhat pregnant) so we had to counter with the extreme brevity of <u>their</u> bathing suits – but a jolly time was had by all during the entire meeting.

--

It might be useful at this point in the narrative to note that the World Mariculture Society expanded its perspectives in 1986 by <u>changing its name</u> to the <u>World Aquaculture Society</u> – as it will be so designated in the rest of this book.

--

Figure 5. Waikiki Beach and Diamond Head on the island of Oahu in Hawaii.

Figure 6. Joan and the author in temporary quarters on Waikiki Beach in Honolulu, Hawaii.

Figure 7. Joan on Waikiki Beach in Hawaii in front of the Royal Hawaiian Hotel in 1977.

Figure 8. Joan and Carl return again to Hawaii's Honolulu Airport to begin his presidential year with the World Mariculture Society at its annual meeting.

Another evening took the form of an elaborate Hawaiian luau with food resembling the early Hawaiian diet – and with elaborate hula dancing groups and with a choir singing Hawaiian songs – undoubtedly the most elaborate demonstration of local heritage that we had ever seen, but typical of the level of expertise that was characteristic of W.M.S. events.

Still another meeting to remember was a much smaller gathering high in the tropical rainforest of the Pali, a mountain near Honolulu. By coincidence, one of the staff members in residence at this "other world" location maintained by the University of Hawaii had been an undergraduate student in my much earlier course at Brandeis University so we spent a great evening, with others, recapturing earlier days, and enjoying the unusual exotic environment that surrounded us.

We did visit, during the years that passed, some of the other Hawaiian islands as well as Oahu, including stays on Maui and the big island of Hawaii, but Oahu and specifically Waikiki Beach remain as our favorites. Go there! You'll never regret it or forget it! We promise you that!

Since this book might possibly be read by younger scientists concerned about their future, it seems appropriate to mention a topic generally referred to as "island fever" and is best described as a concern by young and talented professionals about prolonged service in remote locations (as Hawaii might be considered) and its possible effect on career progress. The federal agency that employed me for most of my career in science has a major research facility on the island of Oahu and several of the youngest professionals with whom I talked some 40 years ago, expressed minor concerns about isolation. My advice at that time was that the scientific community was large enough to encompass minor geographic problems, that it was the science accomplished and the abilities to perform in an outstanding manner that were the critical elements of success in science.

I must have been partially correct in this opinion. One of those concerned junior scientists in that remote location went on to become an outstanding research laboratory director on the mainland, with many substantial publications to his credit and with an exceptional role as a participant in the International Council for the Exploration of the Sea (ICES).

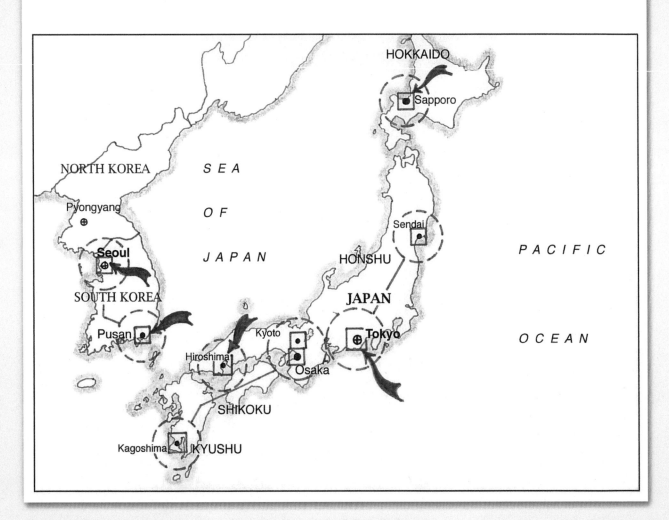

Map 5. Japan and South Korea.

CHAPTER 4
EARLY TRAVELS IN JAPAN

Our first introduction to Japan – a spectacular one – occurred in 1976 at the World Technical Conference on Aquaculture sponsored by the Food and Agriculture Organization (FAO) of the United Nations (UN) in Kyoto, the ancient capital city of Japan. This proved to be a scientific meeting that, after all these years, still stands out in our memories for its excellence in every phase of planning and execution of a technical meeting. This, as it turned out in later years, was typical of the Japanese welcome for visiting scientists, with every aspect of hospitality covered. It was so perfect in every detail that I can't let that meeting slip away in history without some additional accolades.

- The meeting was held in the historic former capital of the country, with its lovingly maintained ancient buildings and shrines, but with thriving modern additions as well.

- Local scientists were out in force and organized to help visiting professionals – far beyond just getting them to their hotel safely, but to be there for the visitors for their entire stay!

- Local scientists were available for advice to visitors who wanted to take the shin-kan-sen – the bullet trains – connecting with distant cities (these high-speed passenger trains operate on their own elevated tracks high above the surrounding countryside).

- Local scientists were available to conduct detailed personal tours of Kyoto – a city replete with historic buildings and shrines – and an admirable venue for a conference with this stature.

- Kyoto has maintained its reputation as a holy city for the Japanese people. It is a city of shrines and classic buildings, as the site of the original capital of Japan.

Following our extremely thorough introduction to Japan at the UN-FAO World Technical Conference on Aquaculture held in Kyoto in 1976 we were released with some concern to the complexities of downtown Tokyo, a universal initial destination with numerous words and notes of advice:

- Like some major cities (New York, for example), <u>taxis dominate the downtown roadways</u>, but they do so in Tokyo from the <u>left side of the road</u> (as they do in London, Sydney, and Brisbane)!

- Subways exist, but may be very confusing; street names may be uncertain but district names are not. Otherwise, use taxis, but have the hotel clerk write a note in Japanese that gives your destination (and a second note for the trip back)!

- After an obligatory tour of the city and its pleasures, get out into the country side – go to Mount Fujiyama, or to Matsushima Bay, or Okinawa, or to any of thousands of attractive tourist destinations beyond the Tokyo city limits!

- Meet people! Learn a few Japanese words of thanks and how to ask about locations of personal facilities, and use those words!

- With some trepidation, reduced by assistance of a Japanese guide, go downtown! We made several forays into downtown Tokyo, to visit typical tourist places like the Ginza, the fish market in early morning, and, of course an obligatory tour of United Nations facilities. (We were housed at the time in a downtown U.S. military hotel, with lots of amenities, even for civilians)! However, I would note that at the time of our first visit (mid 1970's) the U.S. military was a continuing presence in the city!

- If you are a government employee in Japan on official business, by all means try to stay at one of our remaining military hotels in Japan!

Figure 9. Pagoda on the Island of Kyushu, Japan.

Figure 10. Japanese and American members of UJNR on a field trip in Kyushu, Japan.

Since that long-ago 1976 FAO meeting in Kyoto, we have been fortunate enough to have visited almost every island of Japan often for reasonable intervals of time. Our continuing research interests were broad, but somewhat focused on two areas: new methods of shellfish production (raft culture) and disease recognition and control.

Research biologists from the United States and Japan pursued in the late post-war era (1960 – 1970) two approaches to the technology of raising oysters on raised wooden racks or suspended from rafts. Beginning at research facilities at Oxford, Maryland and near Matsushima Bay near the city of Sendai in northern Japan and near the city of Hiroshima in southern Japan, raft culture of oysters – the technology of raft culture of shellfish in shallow coastal waters – quickly resulted in a major food producing industry, and a highly successful one, for Japan. Matsushima Bay is an excellent example of the transformation of a major tourist attraction to the site of a major food producing industry through raft culture of oysters, without losing its place in the tourist handbooks.

The United States and Japan went on to participate in other joint research and development projects through the activities and regular joint meetings of the United States-Japan Joint Panels on Natural Resources (UJNR) throughout my tenure as a government scientist. It was a good concept, and it produced meaningful results! I remain convinced that the singular success of that early (1976) United Nations Technical Conference on Aquaculture held in Kyoto, had long roots in biological research – especially that related to aquaculture – that have persisted to the present time, and have already had effects that have been translated into world food production. The Japanese fish and shellfish production industry is one of the world's best examples of this reality!*

--

One of the unexpected benefits of "late in life" writing (or any writing) is the appearance (sometimes sudden) of "insights" or "solutions" or "answers" to problems of long standing or otherwise. These may not be correct or even feasible, but occasionally they are worth noting.

I am writing now (in this phase of the book) about Japan and the UN –FAO Technical Conference on Aquaculture held in Kyoto, Japan, 1976. I now think <u>that conference contributed significantly to Japan's amazing progress in shellfish production in coastal waters,</u> and I wonder if there is not literature (scientific or otherwise) that would support this conclusion. I have often had questions about the worth of hours spent in such conferences, but in this instance the monetary value of that ever-expanding food industry should be available, as should the total cost of that excellent FAO conference!

--

CHAPTER 5
LATER TRAVELS IN JAPAN

Chapter 4 concentrated on one classic United Nations Technical Conference on Aquaculture held in Kyoto in 1976. I believe firmly that the outstanding success of that Conference may well have influenced the course of research and development in marine sciences, particularly in the development of the fish and shellfish production industry throughout Japan, with which I was involved. This chapter gives me the opportunity to take a broader perspective on ocean research in Japan with viewpoints other than marine aquaculture.

I really want to take a wider view of the world, and I can think of no better place than Japan, with its countless world-class tourist attractions, locations, and distinctive people, to undergo some form of personal transformation from scientist writer to tourist. I (and sometimes my wife) have been virtually hand-carried throughout all the provinces of Japan (concentrating to some extent on coastal points of interest and importance); we have met with countless Japanese counterparts, we have often enjoyed their hospitality, and we have liked almost everything we have encountered there – so what more needs to be said – so much more about people and places!

Japan is of course a nation physically based on a chain of major islands off the Asian coast (from Hokkaido in the north, to Honshu, Shikoku, and Kyushu as we progress southward). The country, to us at least, is best defined by its people rather than its geography. Its people are polite, intelligent, reserved and resourceful, with strong family ties (it must be said, though, that the military history of Japan in World War II belies these statements completely and almost inexplicably, except for the weak rationalization (usually given) that the country and its people were completely enveloped by and in the control of a warlord caste of leaders at the time of World War II)! The United States and Japan are now firm allies in all military matters around the world, and are friendly competitors (if there are such) in world markets.*

--

Most of the readers of this book are undoubtedly too young to have experiences or attitudes about earlier events in the Far East in which Japan was the aggressor. A few may have lasting memories of a defeated, virtually destroyed, but still orderly Japan whose people accepted a new order of life, with a constant view toward proper order and preservation of traditions in the face of radically new social, technical, and political environments. This is the Japan that we encountered!

--

During the time of our repeated visits to Japan (roughly 1960 – 1990) coastal aquaculture expanded enormously. Beginning with a base in shellfish production, Japan also developed fin fish culture industries using a combination of land-based hatcheries and floating sea cages, and shrimp culture, this principally in the southern islands. Development was enhanced in all areas by joint government/ industry programs, with prefectural and national government laboratories as integral participants.

It was easy for me, as a scientific observer and technical participant from another country, during this exciting growth and development period, to be caught up in the enthusiasm and pride of accomplishment to be seen in our Japanese counterparts. They knew that they were deeply involved in what I later began calling "research that matters," and they were justly proud of that fact! To a certain very limited extent, I have always felt that the major fish and shellfish industry of Japan owes a large debt of gratitude for the early research and development activities that we from the United States were involved in so long ago!

Among the many realities that I found professionally satisfying during visits to Japan was the close integration and high status of scientists trained in fish and shellfish pathology (my own chosen discipline). This was due undoubtedly to the large part that fish and shellfish production plays in the country's basic food economy – much more so, for example, than in the United States, where such problems are of relatively minor concern. This scientific synergy concerning fish and shellfish diseases and their population impacts was important to me in many discussions with Japanese scientists, and often provided large areas of common ground in joint meetings. I felt, and I'm confident that the Japanese scientists felt, that there was genuine substance to the exchange – that something of value was being produced by the meeting, beyond the usual paper report. The many results of research reported in the journal "Transactions of the U.S. – Japan Joint Panels on Natural Resources (UJNR)" provide some evidence of the nature and extent of joint research that was accomplished during those exciting times!

Our scientific travel throughout Japan for extended periods of time brought us in close contact with a diverse group of scientists – some of whom we came to classify as friends, and some of whom visited us later here in the United States. A surprising percentage of them had come to this country earlier in their careers for part of their graduate training, and many had reasonable command of English. They, to a person, were immensely proud of the progress that their country was making as a world scientific and economic power. It is not usually customary in Japan to entertain friends and acquaintances within private homes, and we respected this custom, but this was no barrier to development of close personal as well as professional relationships with peers – some of which lasted for many years, and were continuing sources of great satisfaction to us.

I am reasonably sure that no other group of scientists was ever "hand-carried" throughout all the islands of Japan with greater care than was the United States Panel on Aquaculture of the U.S. – Japan Joint Panels on Natural Resources (UJNR). The U.S. Panel members were met individually by Japanese counterparts at the Tokyo airport (Tokyo was always the arrival and departure point for UJNR travel). Technical sessions – scientific paper presentations – and administrative sessions were held in a high-rise building in downtown Tokyo. In the week following the technical and administrative meeting, a joint U.S. – Japan field trip was scheduled for one week to one of the islands. Travel was by air, by the Shin-kan-sen, or by bus to a central prefectural location (usually the capital city of the prefecture). From this center the combined field party visited research laboratories, universities, and large scale aquaculture production operations. The week-long tour concluded on the last day with a series of seminars and discussions of the week's events, before departure from Tokyo.

Needless to say, meetings of UJNR in alternate years in the United States were hard pressed to equal, in quality, those in Japan; but we tried, with the assistance of aquaculture industries, universities, and private research laboratories, to maintain a comparable level of excellence in meeting planning and execution – and in hospitality for our Japanese counterparts. I can affirm readily that of all the foreign interactions in which I have participated during a long career in science, none would exceed the UJNR interaction, and only our participation with European nations in affairs of the International Council for the Exploration of the Sea (ICES) (to be discussed later in the book) would be in any way comparable!

CHAPTER 6
TRAVELS IN SOUTH KOREA

In the early, trembling, post-armistice years of the Democratic Republic of South Korea the United States was, in addition to being a military partner, also a major financial supporter of South Korea – and this condition has prevailed to the present time. One aspect of this support was funding for substantial expansion of growth industries, including shellfish aquaculture, and international assistance in marketing of its products from the sea, in addition to the country's manufactured export trade articles.

One proposal was for direct marketing of fish and shellfish products throughout the world, based in part on acceptance of these raw products by the United States. Another proposal was for U.S. approval of imports of <u>processed</u> seafood – which was more likely to obtain international acceptance. The federal agency to which I belonged became involved in on-the-scene investigations leading to approval or denial of requests (We eventually approved the processed route of international trade – and I was pleased with the outcome.) I had grown to like the South Korean people – hard-working, always pleasant, straightforward, and always pleased to be associated with Americans. South Korea will always remain in my mind as a place populated by gentle but hard-working people living in an equally gentle coastal environment of hills, valleys, and islands, with an unusually productive coastal food-producing industry – all in a usually very mild climate.

Most of my time in South Korea was spent in the southern coastal provinces, but we did make occasional visits to the capital, Seoul, which was to the north and, uncomfortably close to the heavily militarized "demarcation line" with North Korea; a consequence of the still-prevailing "armistice" between the two divided countries. We went to Seoul because many United Nations offices and our own United States Embassy were located there. Even during our brief periods of visitation (in the 1970's and early in the 1980's) and despite the strategic vulnerability of its location, Seoul was busy remaking itself into a modern city, and some of its historical landmarks had already disappeared almost overnight. The transformation of the city of Seoul was amazing to witness. It (the original city) began, on our first visit, as what we expected of an ancient oriental capital, but was quickly modernized (within what seemed like only a few decades) into a westernized capital city, with only scattered remnants remaining of its past glory and history still visible. Our infrequent visits to the "new city" of Seoul were well-monitored by local counterparts in the approval chain for international funding, so these were excellent sources of information about changes in the city's infrastructure since our previous visit – and they were equally upset by the changes to "their" city!

TAIWAN

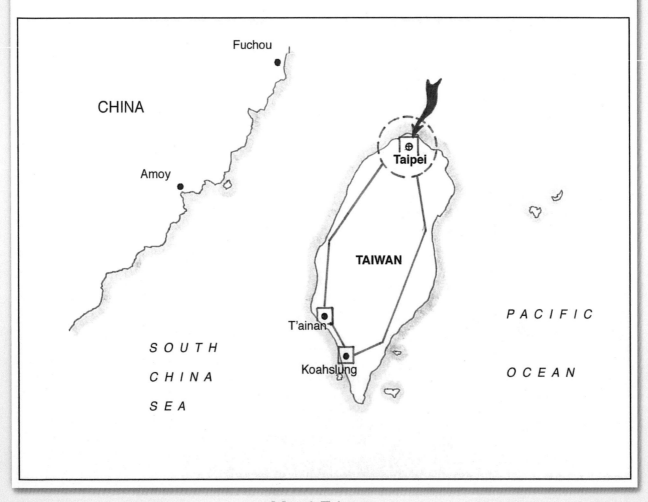

Map 6. Taiwan.

CHAPTER 7
TRAVELS IN TAIWAN

I was standing in the warm, shallow coastal waters of an oyster production farm on the island of Taiwan, 90 miles across a strait from mainland China in 1969. Although I did not fully recognize it at the time, it was a small but important moment in history for me, for Taiwan, and for the United States. I was there on what was an "economic mission" – to observe and inspect and comment on shellfish production methods as well as to offer advice on disease control – all designated to assist the survival and emergence of Taiwan as a country independent of communist China. The real strengths on the island were of course in the tattered aging remnants of Chiang kai-shek's National army still in uniform and still armed, and the presence of U.S. war planes over the 90 mile wide strait that separated Taiwan from mainland China. Representatives of U.S. industries were on the island also, to assist in its economic survival as an independent entity. I was participating in a pivotal moment in world history, and was scarcely aware of it at that time!

With the wisdom that presumably accompanies advancing age, I can now state freely that I and a counterpart director of a U.S. northwestern state shellfish research laboratory were on the island of Taiwan off the coast of China in the late 1960's as "bit players" in a large international drama that surrounded us daily. The nationalist Chinese leader Chiang kai-shek had been defeated on the mainland earlier by communist forces under Mao ze dong, and had retreated across the strait to Taiwan with the remainder of his army. The United States, fortunately or otherwise, was an ally of Chiang and the nationalists.

Invasion of the island of Taiwan, only 90 miles from the Chinese mainland, was at that time considered imminent, but it never happened. U.S. fighter warplanes based in Taiwan patrolled the strait daily during our week-long visit and veterans of the nationalist army, still in uniform and still armed, patrolled the oyster beds of the island that faced the strait.

We were present on that scene and at that time to explore development of a shellfish industry that would help support an independent Taiwan (as were advance parties from a large number of U.S. agencies and corporations – some of which had already set up branches there).

We weren't optimistic. Taiwan is a beautiful island, but its infrastructure consisted mostly of small villages with one major perimeter highway, and two principal cities at opposite ends of the island. Their shellfish research and production facilities were equally rudimentary, although their scientists were enthusiastic about the island's future and cooperated with us fully during our visits to their laboratories.

We made a second visit a few years later. The United States military aircraft presence was muted but still present; some U.S. corporations had opened branches on the island; the circumferential highway had been improved and the anchor cities were larger. Some of Chiang's veterans still worked in the shellfish producing areas – as retirees now, without their weapons. The expected communist invasion never came, but Taiwan had emerged as a successful industrial entity nonetheless!

It should be remembered that the backbone of the Chinese National Army consisted of mercenaries and adventurers willing to risk death at the hands of the communist forces. As was pointed out in the abundant literature of the time, the Chinese Nationalist troops who came across the straits were risk-takers and adventurers – not at all like the communist troops of Mao ze dong, who were largely recruited from the local peasantry, so offspring of Chiang's army (who were the people westerners met on Taiwan) were quite different from the offspring of communist forces on the mainland, with different attitudes toward the western world.

I think that these factors: the presence of the remnants of Chiang kai-shek's army, combined with strong United States support for a rapidly developing economy, were critical elements in the emergence of Taiwan as an independent country. I also think that it is a rare occasion to be on the scene as a participant (however minor) in a major world event – as the survival and emergence of Taiwan was – and as we were.

Taiwan was already beginning to change when we made several extended trips there in the late 1970's from one end of the island to the other. The small towns with thatched roofs and the morning visitations of the "honey carts" were still to be found, as were occasional uniforms of discharged veterans. But there was already a spirit of renewal and energy in the air, fostered clearly by new modern manufacturing buildings as somewhat jarring indicators of progress in an otherwise still rural landscape.

We, as biologists, felt at times like we belonged with the older methods of production on the island, since we were there to encourage expansion of traditional methods as well as to introduce new methods of shellfish production. That segment of Taiwan's industrial expansion was successful, as were many other initiatives by the United States on the tiny island under remarkably stressful conditions of imminent invasion from the Chinese communist mainland, only 90 miles away!

Although I did not return after the 1970's, I have tried to follow Taiwan's successful development on the international stage through various news reports over the years. Its economic success, with great assistance from the United States, is something of a miracle, considering its history, location, and size, and I have always felt good about the miniscule role we played in fostering shellfish trade and research exchange with the United States, in company with efforts of other industries from America! But I also feel, without good evidence, that at least part of the entrepreneurial spirit was infused into the country by the many nationalist Chinese who entered Taiwan with Chiang kai-shek and his army and stayed to be integrated into a new and progressive country. As I have observed elsewhere in the world, it is the <u>spirit</u> of the nation's <u>people</u> that determines to a large extent the nature of a country's future.

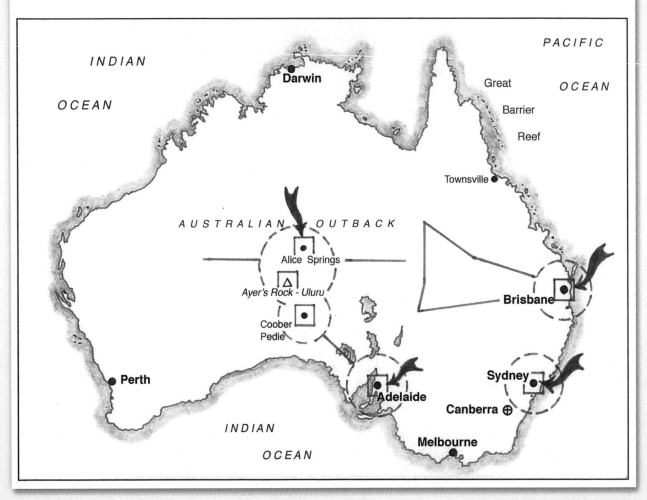

Australia

Map 7. Australia.

CHAPTER 8
TRAVELS IN AUSTRALIA

Thus far in this introductory series of chapters (labelled Phase II: the Oxford (MD) Laboratory and Travel to the Far East) the motivation for the travel has been mostly to supply technical advice to statutory organizations in which the United States was and is a participant. Projected travels to Australia had a totally different purpose: invited participation in a meeting of an international scientific society (International Society of Parasitologists) as session chairman and invited speaker. This invitation was a scientific prize, in my mind, even aside from its attached stipend. Furthermore, it provided opportunity to visit and explore a part of the world – Australia – that had interested me since grammar school days.

The meeting itself was to be held for one week in Brisbane, the capital of the province of Queensland, to be followed by another week for field trips anywhere in Australia. What an opportunity this was – both scientific and global! It was one of those career opportunities that can be dreamed about but rarely realized, and here it was!

Along with planeloads of scientists cum tourists, my wife Joan and I arrived on the continent of Australia in 1984 for scientific as well as personal purposes. The International Society of Parasitologists had invited me to chair a session on marine parasitology and to give a paper at a symposium in Brisbane. This was an invitation that, because of its scientific stature, was easy to accept, especially since all expenses would be paid.

Brisbane turned out to be a modern university city, with a temperate climate and a large welcoming academic population. Most of the academic community seemed to like Americans, or at least to tolerate us gracefully, despite our tendencies to speak a peculiar form of English and to drive occasionally on the wrong sides of their streets.

The technical meeting was well-planned and superbly orchestrated, as most symposia sponsored by international scientific societies usually are. It went on for one week and then it was time for a week-long field trip, with principal choices either the "Great Barrier Reef" or the "Great Australian Outback."* Joan and I chose an intensive investigation of Australia's famous "Outback" and to forego a visit to its equally famous Great Barrier Reef (a dreadful decision to admit in print by a marine scientist, but one aided in part by the fact that Joan and I were both what could be described charitably as "casual swimmers." We both agreed later with the correctness of our decision to forego the Great Barrier Reef expedition, and to instead investigate the "Outback", thereby relegating the Great Barrier Reef and all its colorful inhabitants to the pages of National Geographic Magazine forever. (Forgive us, Jacques Cousteau, if you can, wherever you are!)

--

*The "Great Australian Outback" is hard to describe to anyone who hasn't experienced it. First of all, it seems to occupy much of the central land mass of Australia. Its next characteristic is periodic or long-term dryness, often extensive, with the development of plants adapted to a dry climate and with a fauna equipped for survival in hostile habitats.

A typical outback panorama would include sand, parched soil, scrub bushes, rocks, low hills and very sparse indication of any human presence. Yet people do live here, scattered very thinly in small enclaves and living lives sometimes supported at least in part from outside sources.

To the average outsider (tourist) a question usually arises rather quickly: "Why do people live in the outback and tolerate such extreme environmental conditions?" I certainly have no answers, but then I'm not an Australian. I can only admire and wonder at the persistence, fortitude and energy of those who do populate that harsh habitat of humanity.

--

Beginning in the 1930's, in eighth grade grammar school geography classes in a little town deep in the Berkshire Hills of northwestern Massachusetts, I had developed a fascination for the "Great Australian Outback." I was drawn in particular to the town of Alice Springs – at first because of its "cutsie pie" name but also because of its absolute remoteness. On my early maps, Alice Springs was more than a hundred miles out in the Outback, distant from any other appreciable accumulation of human beings in <u>any</u> direction. According to my early texts it was of great importance as a way station for the early north-south camel trains that carried supplies between Adelaide on the south coast and Darwin on the north coast. The camel trains and their route were predecessors of the famous train called "The Ghan" that is still very much in operation today. The "Ghan" connects Adelaide and Darwin, with a major interconnection in Alice Springs.

Probably of equal or greater importance than its role in transportation, Alice Springs has long served and still serves as the <u>center for electronic communication</u> over much of Australia's forbidding and still sparsely settled central zone known as "The Outback." The Outback consists of untold square miles of sand and scrub and broken hills with rare and isolated communities. The principal contacts for these remote settlements with the outside world were, and often still are, electronic – first radio and more recently television and computer -- with a central station in Alice Springs. This even applies to the <u>Outback educational system</u> with its giant radio (and now television and computer) transmitters located on the edge of town. This provides an extensive and unique support system for each school grade, all across the desolate and endless miles of the Outback.

My early unschooled interest in "The Alice" (as Australians usually refer to Alice Springs) was not unfounded. Located in a geographically unfriendly central part of the nation,* it is indeed "central" and essential to much of the Outback population. But additionally, it has gradually converted itself into a tourist town as well, with a strong indigenous aborigine presence. It offers an array of motels and gift shops, many with art by offspring of the original aborigine inhabitants of the Outback. Alice Springs has preserved the original springs from which the town derived its' name and where generations of camels have been watered. Despite all these recent additions and changes, Alice Springs clearly remains as a place of substance, well beyond the simple tourist requirements, but important to them. For me, it was the realization of a dim boyhood fantasy – and an extremely satisfying one.

From "The Alice" it is only a brief plane ride over the Outback to another favorite tourist gathering place: Australia's unique Uluru Mountains, liberally decorated with indigenous aboriginal art, and with what is Australia's most famous mountain nearby: the enormous symmetrical egg-shaped mountain of stone jutting up out of the Outback, variously labelled the "Red Rock" or "Ayers Rock" or now, officially since 1985, as "Uluru." Its color variations make it an extreme goal for every camera-carrying tourist.

The Uluru Mountains themselves are not like your normal mountains. They come in giant irregular rounded layers of gray rock, as if squeezed from some giant celestial toothpaste tube and hardened in place in the desert. They do, however, form a welcome haven for the average overheated tourist, since they are climbable, photogenic, moist and cool, with many caves with indigenous aborigine drawings.

Joan and I feel qualified for the appellation of "world travelers," and I have tried to summarize some of our experiences in this book. This has been feasible for many of the countries that we have visited but our experiences in Australia have often been so unique, varied, and unusual that I have had problems gathering them in some reasonable order. The foregoing chapter seems robust enough to satisfy the basic purposes of the book – an account of travels as a scientist and administrator – but too many interesting and meaningful experiences have been left untold with the format that I have chosen.

So, for this one country, I have gathered up a few small and large events and places that did not seem to fit the standard format that I have developed for the book. I call this section "Bits and Pieces about Travel in Australia."

--

Alice Springs is located not far from where two major global demarcation lines – the tropic of Capricorn and the 35th line of south latitude intersect near the geographic center of the country.

--

TOURISTS IN SYDNEY

After a crowded but very satisfying two weeks of scientific meetings in Brisbane followed by field trips in the Great Australian Outback, Joan and I descended to the capital city of Sydney on the southeastern coast – where the country had its roots as a British penal colony beginning in 1783. There we were to begin another two weeks of travel that suited us best. Today, Sydney is a modern city, whose residents speak a somewhat aberrant form of English, but where the colonial waterfront buildings still exist as intact representatives of the past, accompanying newer buildings like the new landmark multi-roofed opera-convention-meeting-hall, also on the waterfront, that has already become a symbol and trademark of the city (and the country). We were especially impressed by the extensive waterfront structures that had been constructed back in the days of sailing ships, to process and house the hordes of convicted thieves and other criminals that were released into further bondage almost daily from confinement aboard English sailing ships from London and other ports. This was a story of a nation's founding that is unique in human history. Some of the elements that supported that founding – especially the buildings – are still there!

UNDERGROUND IN COOBER PEDIE

One of the most ambitious forays that we made on southeastern Australia's tourist track was to the city of Adelaide and beyond to the underground <u>town</u>

of Coober Pedie. I wanted to see Adelaide because it was the southern terminus of the original camel trail (and later the train called "the Ghan") that went north through the Outback to Alice Springs and then on to Darwin on the north coast. We also planned to buy an opal in Adelaide which is known as the opal capital of the world, but along the route we fell in with a tour group that was making a side trip to Coober Pedie, a town that is completely underground, having been formed inside a giant opal mine that had operated for decades and then was colonized for living quarters (it even had its own bar and motel and shops). But Joan and I were both uneasy about sleeping underground and we retreated back to Adelaide to buy our opal. We even talked the next day about taking the "Ghan" through the Outback to Alice Springs, but time was getting short on this last phase of our Australian adventure, and by then we felt fully capable of describing the "Australian Outback" to whoever might ask about it, and without a three-day train ride on the "Ghan."

HENLEY ON THE RIVER

On our trip by car back from the Outback we were fortunate enough to pass through a town that was in the middle of a major annual holiday that must be absolutely unique to Australia, where so many rivers dry up completely as part of their annual cycle. Here the holiday is called "Henley Day on the River" and it is so typical of the Australian spirit that I felt that it must be included here.

The holiday actually takes advantage of the dried up river beds that characterize many Australian rivers in dry season. A fleet of <u>racing boats without bottoms</u> is manned by sailors without oars or sails, actually <u>running down dry river beds</u>, accompanied by official boats without bottoms manned by referees and judges (also running madly to keep up). This serious race on a dry river bed is the focus of a daylong shoreside celebration that is properly and traditionally called "Henley Day."

Reading this brief account, or, as we did, actually <u>watching</u> the ritual, and the <u>ebullient spirit</u> displayed during its course, gives a good measure of the individual Australian's view of the world that he or she is fortunate enough to inhabit! So the rivers dry up: what better reason could there be for a celebration? So there is no water in the river bed: we'll make it into an obstacle course with prizes! But, most of all, we'll celebrate our existence here on this good earth!

We left Australia soon after that delightful experience, with the distinct feeling that our sampling of the country was very good, but it was too brief for the immensity and diversity of the place. We will always remember the open welcoming spirit and the constant good humor of every Australian that we met!

PHASE III
THE TROPICAL ATLANTIC
BIOLOGICAL LABORATORY IN MIAMI,
FLORIDA
(1965 – 1970)

PHASE III
THE TROPICAL ATLANTIC BIOLOGICAL LABORATORY IN MIAMI, FLORIDA: HIGH SEAS TUNA AND COASTAL AQUACULTURE RESEARCH, WITH TRAVELS TO AFRICA, THE CARIBBEAN ISLANDS AND COSTA RICA
1965 – 1970

The Tropical Atlantic Biological Laboratory (TABL) on Key Biscayne outside the city of Miami was a Laboratory Director's dream. A relatively new and beautiful waterfront three-story facility with a staff approaching 100, the laboratory was conducting high seas tuna biological research as well as subsidiary oceanographic studies, along with studies of tropical shrimp culture.

This was a large facility with multiple programs and complex interactions with political and academic groups, so it represented a definite administrative growth period for me in a unique urban environment for the family.

Our stay in Miami, in addition to being productive, was pleasant but somewhat stressful for small-town people like us, although we adapted. Joan was able to find and to do superbly at a demanding full time professional job with the well-known newspaper "The Miami Herald." We lived in the up-scale suburb of Coral Gables, where schools were above average. We both faced city traffic every day, but in general we liked our lives in Florida. We were young, with bright kids, and we were "upwardly mobile" (although the thought rarely if ever occurred to us).

In retrospect, this period of our lives spent in Florida was a positive one of professional development for both of us – very stressful, but positive. We survived and prospered, to go on to bigger challenges, as you will discover in later chapters.

Although the research programs of the laboratory shifted in emphasis because of funding and time, this was for me a period of active interaction with foreign counterparts, especially in tuna and tropical shrimp mariculture research.

My functions while director of the Tropical Atlantic Biological Laboratory in Miami were productive and satisfying, and they also provided me with contacts in Central America and Africa – continuing my interest in aspects of international science that I had begun much earlier in Canada's Gulf of Saint Lawrence. I also served for a period during that phase as editor of the journal Fisheries Bulletin, a role that broadened my communication with scientists on national as well as international levels.

In conclusion I would say that my professional experience at the Tropical Atlantic Biological Laboratory was generally a positive one, as were our family experiences, which were wildly variable but also generally positive.

Figure 11. The Tropical Atlantic Biological Laboratory on Virginia Key in Miami, Florida.

COSTA RICA

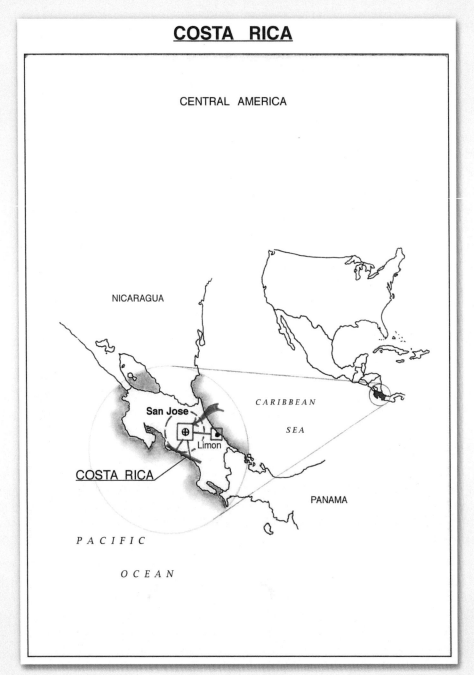

Map 8. Costa Rica.

CHAPTER 9

TRAVELS IN COSTA RICA

During my tenure as director of the Tropical Atlantic Biological Laboratory (TABL) in Miami we hosted a flow of Central American entrepreneurs who were interested in investing in tropical mariculture – principally in shrimp culture in coastal ponds. Many had made preliminary probes into the potential profitability of such ventures but were reluctant because of potential problems such as disease, poor water quality, and inadequate food sources – all biological problems for which research information was urgently needed and not always readily available.

Our laboratory (TABL) was recognized as one source of useful information, through its publications and its senior staff members, some of whom had worked in shrimp biological research for many years and had published extensively. Thus it was entirely logical for members of the laboratory's staff to participate in scientific meetings held in countries with developing tropical marine aquaculture interests – including those in Central America, and particularly in Costa Rica, which had become a center for U.S. foreign investment. One such meeting was that of the World Mariculture Society (WMS) in Costa Rica, widely attended by U.S. scientists with compatible research interests.

Costa Rica, like many tropical and subtropical countries of the world, was eager to participate in the recent boom in coastal aquaculture, especially that concerned with shrimp as a principal crop. World markets for cultivated shrimp have been fed with huge investments, advances in production technology, and a seemingly endless supply of entrepreneurs and investors. Costa Rica, like so many other small, warm, predominantly coastal countries, was an early participant in a major marine aquaculture venture that brought me and the rest of the World Mariculture Society meeting to its shores for an annual week-long meeting, replete with additional field trips to experimental shrimp culture facilities on Atlantic and Pacific coasts of the country. The meeting, like most of those of the World Mariculture Society, was a huge technological as well as social success, and Costa Rica rapidly assumed a role as player in the shrimp aquaculture business soon thereafter.

In Costa Rica, as well as in other developing countries around the world we had the opportunity to visit, the scientific research relationship involved not only scientists but also business and investment partners, and it was capable of having a significant impact on the economic as well as the technological development of the country! But our extended visit to Costa Rica was not limited to scientific society sessions. We as scientists and society members enjoyed the field trips especially – with complementary visits to experimental facilities on both coasts of the country, and with the best of amenities provided.* Some of us stayed over for additional travel in a truly beautiful and compact nation. (It even provided an active volcano for tourists at the time, as I recall).

The period 1965 – 1980 turned out to be one of intense interest in bioactive substances from the sea, and on the potential cultivation of animals producing bioactive materials (Kaul and Sindermann, 1978). The research interest of the period was positively reinforced by investments of the pharmaceutical companies in biochemical studies of many marine organisms. Industry support of the World Mariculture meeting in Costa Rica was also a reflection of a positive assessment of the success of those research approaches, even at an early stage. The scientific research relationship involved not only scientists, but also business and investment interests with a corresponding significant impact on the commercial as well as the scientific development of Costa Rica.

--

There are, of course, certain risks involved in any travel in less developed parts of the world, and this reality was impressed on us, vicariously after the Costa Rica meeting was over, when we all received bad news of a very unsettling episode on the so-called transcontinental highway through Mexico, when two well-liked acquaintances from the World Mariculture Society who elected to drive to the meeting in Costa Rica <u>disappeared</u>, along with their car, and no trace was ever found – to my knowledge – of them or their car!

--

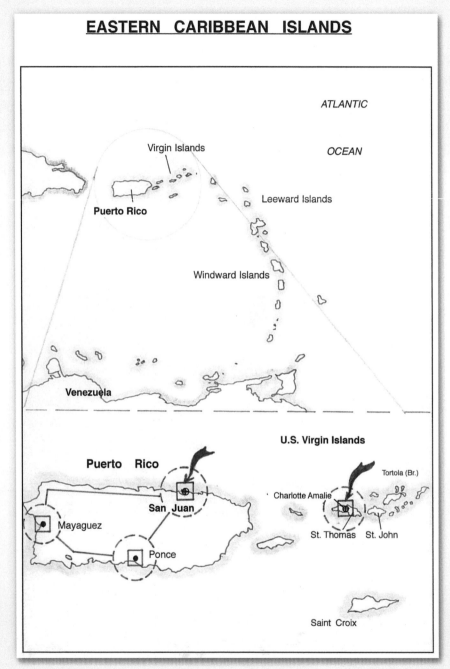

Map 9. Eastern Caribbean.

CHAPTER 10
CARIBBEAN ISLANDS: PUERTO RICO AND THE U.S. VIRGIN ISLANDS

During my tenure as Laboratory Director of the Tropical Atlantic Biological Laboratory in Miami, we made a number of trips to the Caribbean Islands, particularly to Charlotte Amalie in the U.S. Virgin Islands and to San Juan in Puerto Rico. At that time Chalk Airlines was offering PBY seaplane service* from its Miami seaplane base to the islands, and that is how we travelled. Our youngest son Carl was about 5 years old during that period, and he got to be an honorary co-pilot on one of those beautiful and to us historic seaplane flights to the islands.* Our visits to Puerto Rico were typically as tourists, except for an occasional scientific conference, usually with residence in a hotel in Old Town San Juan, and trips by car to points of interest around the island and the beaches, the radio telescope station, the rain forest, and the luminescent bay on the west coast. I remember clearly one evening affair that included a visit to the western city of Mayaguez, specifically to see that natural phenomenon of bioluminescence on a large scale, mainly because of a wild bus ride on the narrow curving roads with a drunken contract driver, until a member of the university faculty hosting the field trip mercifully took the wheel. (I was a great cigar smoker at that time, and Charlotte Amalie, on the island of St. Thomas in the U.S. Virgin Islands, had shops with the best selection that I have ever seen).

We visited other large scale places and events in Puerto Rico too. One that impressed me most of all I think, was the communication project in which an entire natural valley had been equipped as a giant radio telescope, sending out news (signals) of man's existence on earth to the rest of the universe. I asked if we had had received any replies, but nobody seemed to have an answer (or maybe I just asked the wrong people). At any rate, the sheer size of the operation, engulfing and modifying an entire mountain valley for scientific purposes, was overwhelming enough for me, and I went away happy with human ingenuity in what I saw (until some time later, when I read that Republican politicians had forced budget cuts that had reduced the federal operating funds for this beautiful innovative research program! For shame!).

--

As I recall it, the seaplanes that were in use by Chalk Airlines were surplus navy Grumman PBY patrol planes converted for passenger use, and were original sources of the old Navy Pacific battle cry of World War II, "Don't send me out to die in a g— damned PBY" (referring to the relative slowness of the PBY and hence its vulnerability to destruction by Japanese fighter planes). But they were great for our purposes, and could be boarded at the waterside in Miami.

--

One of my most persistent memories of Charlotte Amalie in the Virgin Islands (other than the purchase in a downtown shop of the sick metal fish sculpture pictured on the cover of this book) was the overpowering physical presence, parked by the main road to the Charlotte Amalie airport, of a giant airplane made <u>entirely of wood</u>. This, apparently was the creation of the early 20th century eccentric millionaire and entrepreneur, Howard Hughes, and was dubbed the "Spruce Goose." It was presumably, at one time, in flyable condition, but was now parked, hugely, by the roadside, engineless (although, as I recall it, there were pods for eight engines on the wings). It was genuinely massive, but, unfortunately, we had no time to investigate reasons for its presence there. Had it actually been <u>flown</u> there, or simply <u>built</u> there? My computer will tell me if I ask the right question, but even computers may reject answers that are too trivial!

At any rate, we found the Virgin Islands and Puerto Rico to be pleasant and accessible destinations for occasional scientific meetings as well as for their tourist appeal.

Map 10. Western Africa.

CHAPTER 11
TRAVELS IN AFRICA

Our sojourns in Africa occurred mostly during my tenure as director of the Tropical Atlantic Biological Laboratory in Miami (1965-1970). Joan and I first went to Africa as tourists to Tangier in Morocco. Later, I went as scientific advisor to several meetings of the UN-FAO Commission for the Exploitation of the Central African Fisheries (CECAF) in Accra, Ghana and Casablanca, Morocco. We explored no jungles, and saw no tigers, but we did see something of central coastal African cities and their people, as well as much of the countryside, by various means of ground transportation as well as extensive air transportation.

Our sample size was obviously too small to develop any valid judgments about continental Africa, except possibly to observe that a selected few of the former colonial powers, and most exceptionally the United Nations, seem to be providing much needed infrastructural support for an extraordinarily weak governance for much of the continent. Some former colonial powers – notably France – were still involved (in supportive roles) in needed places and functions – of which there seemed to be many!

Several of the major cities of Africa now have giant United Nations Meeting Buildings. Accra, the capital of Ghana, is one such location. The building is an office building but is also a center for large regional U.N. committee meetings. It is of circular design, with a large meeting hall with concentric rows of seats flanking the core of desks for active participants, and a flanking outer circular zone of interpreters' booths – all with elaborate sound and visual projection equipment – all interwoven with thick red carpeting. The meeting I attended in June 1969 of the United Nations Commission for the Exploitation of the Central African Fisheries (CECAF), was one of a series of regular semiannual meetings. I was there as a scientific advisor to the United States delegation.

We were, temporarily, a tiny part of the United Nations' effort to develop a viable economic infrastructure for some of the African nations. In our specific case it was to aid in the development of effective long-term management of the abundant marine fisheries resources off the central west African coast – an important world resource, and one with which the United Nations was and is very concerned.

We visited Africa at a time when distant water fleets of factory fishing trawlers from many countries, including the Soviet Union, Poland, East Germany, and other European countries, were exploiting the abundant fishery resources of the Central African coast, especially in the Gulf of Guinea, to the detriment of local fishermen. It was a great time for scientists to take their proper places as informed advisors on a world stage, and those of us there in that capacity at that moment in history felt good about our role as a source of reasoned comment. We were there, in the right world forum, with at least some of the right data, and the role of science was clearly defined, even though we knew that an effective international legal infrastructure did not exist to support the necessary proposed catch limitations.

Actually, though such rigidly controlled UN-FAO meetings as those in Africa offered little opportunity for exchanges or discussions with <u>local</u> scientists, they did offer many opportunities for discussions with scientists from <u>technically advanced countries of the world who were also there as technical advisors under UN sponsorship</u>. This reality made every UN-sponsored meeting very worthwhile, regardless of its location, since cadres of excellent international scientists from many countries were assembled by the many Commissions of the United Nations over the years, and these professionals take and have taken significant roles in discussions that lead to international actions by UN-FAO!

Growing up as I did in the relative wilderness of Western Massachusetts early in the last century, I had a somewhat idealized concept of Morocco and much of the adjacent territory as a hot and sandy place where movies were made ("Road to Morocco" with Bob Hope and Bing Crosby, "Casablanca" with Lauren Bacall and Humphrey Bogart) and people still rode around over the desert on camels and haggled in tiny crowded city shops. Just the names of cities: Casablanca, Fez, Marrakech, can bring back vivid memories of my earlier movie impressions of that part of Africa – without ever having been there previously of course.

In my later career as a peripatetic scientific research laboratory director with the federal government I achieved the unlikely: visits to that part of western Africa that was almost familiar to me because of those early movies! A city with a familiar name (from the movie of the same name (Casablanca) was to be the site of another UN-FAO meeting which I was to attend as a scientific advisor. We met formally in Casablanca in another large circular conference building (like many UN-FAO Conference Centers in other African countries and around the world – elaborate and impressive, as befits the world organization.

We visiting scientific advisors were "short-term people," so we were housed in modern western style hotels in downtown Casablanca, with regular bus service to the meetings – service that passed through parts of another world of African big city life, with constant crowds everywhere.

We did, though, arrange a free day for a group bus trip to Marrakech and its fabled Medina – an entire section of the city with narrow streets and literally thousands of small open-faced shops selling everything that is saleable in the world. We bought rugs, robes, cigars, anything that appealed to us, and we temporarily lost sight of our local guide, at which point we were set upon by a so-called "children's gang" of grubby noisy kids snatching at clothes or anything vulnerable. One of our party members – a "Brit" – lost his wallet and passport before the gang was dispersed, but other than that incident, the day was a huge success, and we escaped from the Medina tired but satisfied! After this brief encounter with the real world of Africa, most of us were ready to return to the elaborate surroundings of the UN-FAO building and western-based hotels the next day.

The cities and larger towns of tropical and temperate western coastal Africa that we visited – places like Accra, Marrakech, Fez, Casablanca, and Tangier – could all be squeezed into a master mold of ancient, hot, overcrowded, and "uneasy" places to visit, without some over-riding objective (as we had). Cities exist where westerners are advised to travel in organized groups and not to venture outdoors alone at night. Life for many local citizens seems to be harsh and almost a daily struggle for survival. Some tourists can ignore this component of African existence, but I am not one of them. I do admire, though, the massive efforts of international organizations like the United Nations, and even efforts by industrialized countries like United States, France, and certain others, to help these people to bring some order from chaos and corruption.

With so many other brighter pleasanter destinations in the world, my advice to tourists would be to go to Africa, as we did, only as part of some organized international group – private or governmental – with a specific mission to accomplish and with offices or other facilities on that beleaguered continent – but to go and work and observe and return with a clear understanding that most African problems that you will encounter seem to approach the insoluble. (and I have never felt that way about any other part of the world that we have visited)!*

--

As a final note for readers who are also scientists, I would point out that here and there in the world, and especially on the West African Coast, are remnants of a once proud French colonial scientific organization – ORSTOM – the Office for Research in Science and Technology in Distant Seas (very roughly translated), which, like its more famous counterpart, the French Foreign Legion, has refused to disappear with colonization but has undergone metamorphosis to become a support and assistance organization for formerly French colonies. My contacts with the scientists of ORSTOM have been few but positive, and the concept supporting their continued existence and support function should be appreciated worldwide!

--

PHASE IV
THE SANDY HOOK, NEW JERSEY
MIDDLE ATLANTIC COASTAL
FISHERIES RESEARCH CENTER
(1970 – 1990)

PHASE IV
MARINE POLLUTION AND FISH DISEASE STUDIES AT THE MIDDLE ATLANTIC COASTAL FISHERIES RESEARCH CENTER, SANDY HOOK, NEW JERSEY
1970 - 1990

The move to become Center Director of the Middle Atlantic Coastal Fisheries Research Center of the United States Department of Commerce at Sandy Hook, New Jersey, represented achievement of my ultimate goal in federal fisheries research and administration. My administrative responsibilities were to be broad, encompassing all the federal fisheries research laboratories on the entire Middle Atlantic coastline of the United States from Connecticut to Virginia, but my physical office location was to be in a marine research laboratory – the Sandy Hook Laboratory – providing a research environment for its location. This was a position in which I could remain actively involved in ongoing research.

Our Center research programs were broad, covering shellfish as well as fin fish, and including environmental problems of pollution, disease, and over-fishing. Shellfish aquaculture was also an active research area. Communications with European colleagues increased, especially in areas such as concerns about effects of coastal pollution and the transfers of non-native marine species to the United States. These activities involved frequent travel, often to foreign countries for meetings with international groups such as ICES. I was appointed chairman of the ICES Working Group on Introduction of Non-Indigenous Marine Species, for a ten-year period from 1980 to 1990. I remained in a research laboratory environment, yet I was administratively responsible for the research of all federal fisheries research laboratories in a large sector of the nation's Atlantic coastline – all in areas of marine studies that were of most interest to me: disease effects on fish and shellfish, population studies of marine fish, pollution effects on fish and shellfish, and marine aquaculture.

Professional members of the separate laboratories worked well together, and progress in research was excellent, in my opinion. The complex geographic structure was challenging, to say the least, and it required extreme amounts of inter-laboratory communication efforts, but research productivity (the ultimate criterion of success) was good. I was subsequently awarded a U.S. Department of Commerce Silver Medal for administration of geographically dispersed research laboratories.

Our family position in New Jersey was pleasant. We lived in the small town of Fair Haven, near but not on the northern coast. The schools were good for our youngest son Carl, and the older kids were in college or already out in the world. Joan took another demanding professional job in publishing, this time with a book publisher in the nearby city of Red Bank, where she quickly became an integral part of the complex process of book production there. During most of this period we owned two homes (one in New Jersey and one in Miami) so that we and all the family members could enjoy intervals of sub-tropical living during scarce (for us) vacation periods. Those years were good and expansionist but stressful ones for us!

In retrospect – and this is time for a little of that – our family has lived, for different periods, in many choice locations on the Atlantic Coast of the United States. We started out in Cambridge, Massachusetts while I was at Harvard and then teaching at nearby Brandeis University. We then all moved to Boothbay Harbor, Maine for nine years, while the older kids were in elementary and high schools. We then moved to Oxford, on Maryland's Eastern Shore when I was offered the position of Laboratory Director of the Oxford Biological Laboratory. After five years there I became Director of the Tropical Atlantic Biological Laboratory in Miami, Florida, where some of the younger kids were in elementary and high school and the older ones were in colleges. In 1970 we moved northward again to New Jersey and the Middle Atlantic Coastal Fisheries Research Center at Sandy Hook. This was to be my final and most challenging research directorship with the national fisheries agency – a career choice that I made so many years ago and never regretted!

CHAPTER 12
TRAVELS IN ICELAND

During the period from 1970 to 1990, when I was making annual or more frequent visits to Europe, I usually travelled by American airline companies that made no stops in Iceland. Sometime during that period I decided to book the flight to Europe on Icelandic Air that did have a convenient major stop in Reykjavik, and the price was right. We arranged for a stopover in Reykjavik and a one day circumferential tour of the whole country by car. The tour permitted convenient stops (at our discretion) to see an active volcano (in the news at the time), an active whaling station processing bloody remains of a whale, a thermal hot water distribution station, as well as much subdued vegetation, where there are no trees but lots of moss and other low-lying vegetation.

I would not recommend a one-day tour of <u>any</u> country, but I did feel that we at least had some basis for recognition when the country's name was mentioned – and we had previously met a number of very gregarious and well-informed Icelander scientists at ICES meetings. We even had time (on a second day) to explore downtown Reykjavik at night. The city was then just emerging as a modern place; aided to a great extent by a universal underground thermal heating system and a booming oceanic fisheries industry.

A few years later we did actually attend an ICES meeting in downtown Reykjavik, and were amazed at the development of the city. Its fisheries and ship-building base had expanded significantly, with aluminum and other heavy industries contributing to the economy, and with the nightlife booming as before. It was a great stopover in an unlikely location!

CHAPTER 13
TRAVELS IN WESTERN EUROPE

Western Europe is almost too diverse, too complex, and too interesting to be considered as a single geographic entity in this book, especially since Joan and I made so many prolonged visits to so many legal and ethnic divisions of that geographic zone over an extended period of years. We probably spent more time in Western European countries than in the aggregate of all other countries in the world (except, of course, the United States), and enjoyed our time there immensely.

Just as an example, we made several extended visits to northern Scotland, including stays in Aberdeen and visits to the north and west coasts (Isles of Hebrides). We then spent a week in London, officially at a symposium on "Oil Pollution in the Ocean" sponsored by the Royal Society of London, who very thoughtfully supplied adequate time for sightseeing in the city. We also made an extended but similar visit to Dublin, Ireland, a few years later, following a meeting there of the ICES Working Group on Introductions of Non-Native Species – a meeting which was followed by an extensive tour of the dramatic southeast coast of Ireland, visiting places like Cork, Bantry, and Killarny.*

We made many extended visits to other Western European countries, for ICES meetings, for FAO-UN meetings, or for international scientific society meetings, but some of these can be best described country by country in following chapters, in the interest of possible greater clarity, according to the following alphabetical listing:

An informal, undated, and alphabetical listing of our travels in Western Europe, of variable lengths and some with multiple repetitions, is included here:

- Britain (United Kingdom): England, Scotland, London, Edinburgh, Aberdeen, Hebrides.

- France: Paris, Nantes, Avignon, Arles, Aix, Brittany, Normandy, Provence, Marseilles, Nice.

- Germany: Hamburg, Munich.

- Ireland: Dublin, Cork, Killarney.

- Italy: Rome, Venice, Verona.

- Low Countries   .

- The Netherlands: Amsterdam, The Hague, IJmuiden.

- Belgium: Arlon, Bastogne.

- Portugal: Lisbon, Porto, Vigo.

- Scandinavian Countries.

- Denmark: Copenhagen.

- Sweden: Göteborg, Stockholm.

- Norway: Oslo, Bergen.

- Spain: Madrid, Salamanca, Santiago de Compostela, Santander.

What emerges from our extensive travels in Western Europe is a genuine potpourri of impressions – almost all favorable – from the airport greetings to the guided tours to favorite places to visit to locations of historic or scientific interest or gustatory prominence. Western Europe was the part of the world we visited often enough and for long enough periods of time to have formed good feelings about – almost regardless of national borders (although we had some favorites as I have already admitted).

--

Some language purists may disagree with my inclusion of the British Isles and Ireland as "Western Europe" but to a simple tourist like me these islands form part of a geographic zone that should be labelled "Western Europe" – as I have done here! My listing of Western Europe's principal subdivisions that we visited (strictly for tourist purposes) would be as follows at the end of this chapter.

--

NOTE: WESTERN EUROPE AFTER WORLD WAR II

During the decades immediately following the end of World War II – well before our first visits – many cities and towns of Western Europe were dismal ruined places to visit for any purpose. Reconstruction of war damage was slow and a pervading sense of gloom and disorder was common among survivors. Restoring the political, social, and physical infrastructure of towns and cities was considered a worthwhile objective to be supported by occupying forces. Such reconstitution included schools, colleges, and universities, all of which had suffered loss of faculties, equipment and funding. Far-sighted programs such as the United States Marshall Plan did much to restore a semblance of order and function to war-affected countries in Western Europe, especially in the early post-war years, when many peoples' lives remained in poverty, and chaos was ever-present, every day and everywhere.

Full implementation of the Marshall Plan of Economic Aid from the United States to Western Europe after World War II was probably the most far-reaching economic action, as measured by long-term results, of any economic steps ever taken there (followed only by the more recent development of the European

Community). The United States played a major role in the revival of Western Europe following World War II!

Most of the scientific travels reported in this book were conducted after 1965, when much of the world had returned to a large degree of "uneasy normalcy" – when people had picked up pieces of lives and careers that had been disrupted by conflict. Many of the scientists whom I have included in this and previous books were, like me, touched or basically affected by the war and its aftermath, mostly negatively but occasionally positively. (My decision, for example, to become an ocean research biologist instead of a physician – one of the best career decisions that I ever made – was based almost entirely on my experiences as an enlisted infantry platoon medic during most of World War II in Europe). I have never regretted that decision!

So, with these introductory comments, I invite you to join us briefly in some of our travels in Western Europe, mostly in Phase IV of my scientific career while I was Director of the Middle Atlantic Coastal Fisheries Research Center at Sandy Hook, New Jersey.

SPAIN AND PORTUGAL

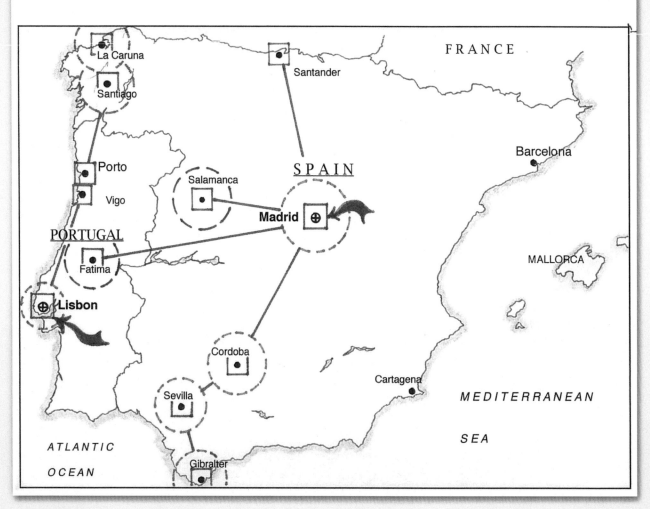

Map 11. Spain and Portugal.

CHAPTER 14
TRAVELS IN SPAIN AND PORTUGAL

Early in our European travels, we had the extreme good fortune, during a visit to Spain and Portugal, to have a scientific meeting at La Caruna on the northeast coast of Spain. We chose to stay in the town of Santiago de Compostela, a town rich in Catholic religious and Spanish Royal history. We enjoyed this trip so much that we scheduled a later commercial bus trip throughout the interior of Spain, to augment future periodic visits to coastal locations. Some of our observations and comments about this delightful country are on the following pages.

"Santiago de Compostela" on an early summer morning – words that bring back pleasant memories of a visit to a fabled coastal town in northwest Spain, filled with well-preserved and restored remnants of its royal and religious past. We were staying in a small guest hotel named "The Five Kings" across the enormous cobbled town square from an ancient Catholic cathedral that housed the crypts of those kings. The cathedral was the terminus of a famous pilgrims' and penitents' pathway from Madrid to Santiago de Compostela, called "The Way of Saint James."

Our huge hotel room (suitable for a king) looked out over the plain, where, on the road from Madrid on that early morning was a procession of long, long lines of pilgrims and penitents – hundreds upon hundreds of them – walking the famous Trail of Saint James, from Madrid to Santiago de Compostela. It was one of those moments in time that, if witnessed, can never be forgotten, and I remember it as clearly as if it were yesterday.

In Spain we were able later to visit some of its <u>internal</u> points of interest by going on commercial bus tours. For example, we spent some time in the University city of Salamanca, even visiting some of the ancient university buildings and sitting in ancient classrooms. We visited museums in Madrid, and attended a bullfight (not very happily) at the insistence of accompanying friends on the tour. Mostly, on these inland bus tours, I was impressed by the quantity of very visible remnants of ancient Roman stone-works throughout the country. Also impressive were the exquisite and well-preserved examples of Moorish palaces and gardens.

We were reasonably pleased with our tour bus venture but we decided that we much preferred our usual bumbling relatively loosely-scheduled but carefully planned approach to long distance travel abroad by rental car, preferably with good friends. This, in retrospect, is the way we should have thought about this bus tour venture – and an observation that we offer to future travelers. For us, the degree of independence would be a determining factor in most decision-making on this important issue.

Figure 12. The overwhelming gates to the city of Avila, Spain in 1975.

NOTE: Joan and I have problems with European bus tours (or any bus tours) –even the more expensive ones…The most serious problem is with human physiology, which is not immediately adaptable to long periods of total inaction interspersed with short intervals of frantic activity centered around satisfaction of bodily functions.

We have found a much more satisfactory method of transport: <u>a rental car</u>. Of course a rental car cannot be considered a panacea either, especially in countries where drivers persist in driving on the wrong (left) side of the street – as they do constantly in the U.K., Ireland, Japan, and too many other places in the world. Those of us from America with proper right lane reflexes are at constant risk if we drive rental cars in those recalcitrant countries, as I can affirm after close encounters on some of their narrow roads. But, if you can't take the risk, take the bus!

Other advantages of rental car use are that the route you take and the stops you make are completely at your discretion (thus requiring substantial advance planning). For this I highly recommend your computer for detailed information on some of the most remote locations or subjects that might be of interest (like Findhorn, Scotland or Martello Towers), as well as for any of the more popular destinations in any country. We discovered quickly that the amount of information available this way is useful in planning any trip. You then go as an "informed traveler" rather than a "casual tourist," and that makes a worthwhile difference – at the time of the visit and subsequent to it. (This is one of the important realities that we have confirmed in more recent travels to Avignon in France.)

Our visits to Spain were always overwhelming, whether we traveled by bus or car – principally for the almost continuous vast vistas of stunning natural scenery superimposed on a (to us) surprising amount of the remnants of the Roman Empire – some in remarkably intact condition. I think this was the most impressive aspect of the landscape, for us. Of course, throughout Spain are palaces of antiquity always accompanied by lush gardens of stunning beauty. The cities that we visited, Madrid, Salamanca, and many others, were exciting, if challenging, for us as tourists.

PORTUGAL

Our visits to Portugal were enhanced by its proximity to northwestern Spain and the town of Santiago de Compostela where we made several visits. But our visits to Portugal itself were scientifically induced, since the country at the time was undergoing remarkable expansions of its fisheries research and development programs, especially its molluscan shellfish (oyster and mussel) programs. We spent several very pleasant and productive days on one visit in the hands of the Director of the Fisheries Research Laboratory at Vigo. He was an energetic enthusiastic person with an exquisite sense of humor, who spent those days with our visiting group of ICES scientists, traveling up and down the coast of Portugal, displaying and explaining with great pride the expanding fishing industry of the country, and teaching all of us the correct pronunciation of the island of Tenerife (with emphasis on the final "e"). Western Europe seems crowded with such educated, aware, and outgoing people like him!

Our entrance into Portugal from Spain brought us quickly to the thriving religious shrine town of Fatima, with a world-wide reputation for miraculous healing powers of its Shrine of Our Lady of Fatima. We offered our individual masses for silently stated purposes for ourselves and our family, but we then realized that many petitioners in the church with us had enhanced their requests for divine intervention by <u>bringing with them plastic reproductions of body parts</u> that <u>were to be subjects of miraculous reclamation</u> – arms, legs, brains, kidneys, lungs – a whole panoply of plastic body parts which they then deposited in appropriate bins in the front of the chapel. We had noticed, but had given only slight attention to lines of shops along the street leading to the shrine, selling various souvenirs, but in particular these plastic reproduction body parts (I still have my plastic leg bought as insurance against falls getting on and off our tour bus).

We made two extensive visits to Portugal, one from the east after our tour of inland Spain, and one from the north after our visit to Santiago de Compostela. Our probe from the east took us to the capital city of Lisbon and then to Genoa, while our probe from the north took us to the coastal cities of Porto and Vigo where we were treated royally by members of the Portuguese National Fishery Agency, who arranged for us elaborate tours of emerging coastal shellfish development programs. In the course of these associations, I came to an interesting finding: the Portuguese (except for their language) are in many ways much like the Irish, at least in their approach to life and living – easygoing, friendly, and always ready for a good time. They (the Portuguese Fisheries Research people) took us in hand for four days of demonstrations, travel, and good fellowship that we enjoyed thoroughly, and will long remember.

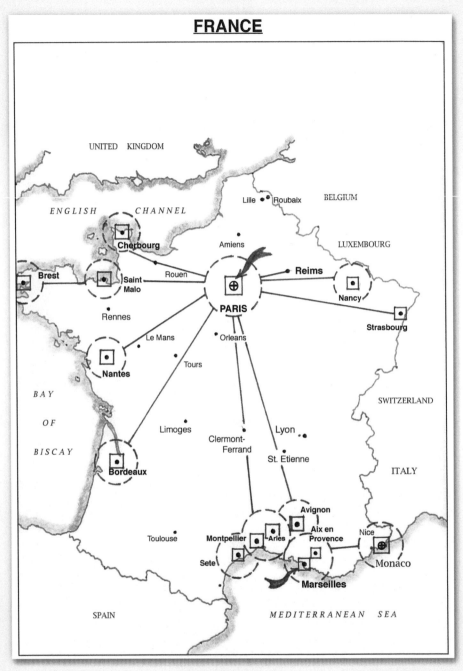

Map 12. France.

CHAPTER 15
TRAVELS IN FRANCE

A Forword to this Chapter

France, one of our most favorite countries of Western Europe, is also extraordinarily complex geographically, with coastlines on the Atlantic Ocean and the Mediterranean Sea; with natural borders with Spain, Switzerland, Germany, Luxembourg, and Belgium; and with an extremely complex land mass from coastal plains to mountains with vineyards everywhere.

Think of France as a badly-battered but somehow welcoming giant compass face, with Paris as a somewhat northerly displaced <u>hub</u> for the compass needle, but always the <u>hub</u>. Thus to go to the Normandy or Brittany Peninsulas, go west from the hub; to go to Nantes, or Bordeaux, and the Atlantic Coast go southwest from the hub; to go to Provence, Nice and Marseille go south from the hub, and to go to Alsace-Lorraine and the low countries go generally east and northeast from the hub.

With these general compass guidelines and a good tourist map it is easy to chart a journey of days, or weeks, or months if you have the time and money – but always returning to Paris as the hub, for renewal and further adventures there in that fabulous city. (Of course the time and place of much of <u>our</u> travels were related to scientific or administrative priorities, and were rarely ours to choose, but even with those limitations we were generally satisfied with the destination and the season of the year).

Figure 13. Joan confronts the Eiffel Tower for the first time (1958).

FRANCE

France really has almost too much of everything that any tourist, even the scientist-tourist, would appreciate. Beginning in the hub, Paris, a leisurely day at one of the finest museums in the world, the Louvre, is certainly in order. Scientists then need to pay homage to the world famous Sorbonne, with its associated medical school and university. Then there are structures unique in the world, like the Eiffel Tower and the Cathedral of Notre Dame, and many places of overpowering beauty, like the Tuileries Gardens. There are also countless small places of great charm, like the tiny west bank hotel on the Rue de la Harpe that welcomed us on every visit, or the small pub near the Champs d'Elysée where we spent many a pleasant hour.

PROVENCE

Moving out of Paris as our hub, we went south to Provence where we took a week's vacation in a small country hotel with a large pool just outside a small hilltop village near Avignon. We divided our time, between the provençal sun by the pool and day trips to famous cities from medieval times: Avignon, Aix, and Arles, for a week-long completely pleasurable visit that we will never forget.

We made another, briefer, visit to Provence, when we emphasized return visits to the larger cities, following an ICES Working Group meeting in Sête. These cities of Provence are clearly noted for long ago events as evidenced by remaining ancient structures of historical interest in every one of them, but they also seem to have a vigorous modern overlay to the antiquity which makes them much more interesting to the average visitor. To have this much relevant antiquity in cities with adequate tourist amenities is surely a highly favorable factor, especially when such cities are located in areas like Provence with still other tourist-friendly features of climate, country-side, and small villages with easy access to the major cities. However, in the matter of the cities of Provence, I strongly recommend some brief preliminary historical background studies before a visit, especially for Avignon. (And do as we did on our first visit: book a small hotel with pool in a small town nearby).

NORMANDY AND BRITTANY

On a subsequent trip to France in the late 1970's Joan and I went west from our arrival hub, Paris, by rental car to the Normandy Peninsula, mostly to see how things had changed since the dark days of 1944, after I had landed there two weeks after D-Day as an infantry platoon medic. We were immediately very pleased with the transformation back to what must have been its former quiet orderly countryside, with rebuilt towns and cities and wider roads replacing the awfulness of war ruins – with, in some cases, completely rebuilt towns where only ruin and desolation had existed. War memorials and cemeteries were abundant, but now it was the spring of a much later time when apple blossoms were blooming, crops were being planted, and the world itself was so much different from what it was on my earlier visit!

We continued on to the Brittany Peninsula and to the French Oceanographic Institute in Brest, where I stumbled through an impromptu lecture (described as a vignette elsewhere in this book). After that debacle we continued down the French Atlantic Coast to the city of Nantes, where a major laboratory of the Institut Scientifique et Technique des Pêches Marine (ISTPM) is located, and where the laboratory director was a good friend and an associate in the ICES subgroup which was to meet at his facility. The meeting went well and we all enjoyed the exceptional hospitality of his stunning coastal home before departure.

NICE, THE CÔTE D'AZUR (AND TWO HOURS IN MONACO TOO)

One of our best visits to France was related to a meeting of the ICES Working Group on Introductions and Transfers of Marine Species, held this time in Sête on the Mediterranean Coast near Montpellier, and the location of a French

Institut Scientifique et Technique des Pêches Marine (ISTPM) Shellfish Research Laboratory. After the ICES meeting, which dealt in part with the problem of an introduced invasive alga that was spreading around the Mediterranean Sea, Joan and I rented a car to explore the nearby and famous south coast of France, including the Côte d'Azur, Cannes, Nice, and on into the principality of Monaco, which is immediately adjacent, and has a world-famous marine laboratory, aquarium, and casino that we wanted to see. The laboratory at the time was under directorship of a well-known public environmental figure, Jacques Cousteau, and the facility at the time was embroiled in scientific controversy about its possible responsibility for introducing the harmful alga to the Mediterranean Sea (and eventually to California coastal waters). The aquarium outflow pipe was a suspect. I was an interested observer only, at the time, but I remember clearly that the marine laboratory and its associated aquarium were atop a hundred foot high headland above the Mediterranean, and my first private question was, "How in hell do they get sea water up there anyway?" Most of my questions remained unanswered because we had earlier used up our available time in Monaco at the Casino de Monte Carlo, which was conveniently next door to the Laboratory and Aquarium (We had elected to take a tour bus from Nice rather than to drive our rental car). However, on the trip back to Nice we did stop in the town of Grasse which is world famous for its blends of perfumes. Joan bought enough, mostly in tiny vials, to last her and our daughters Jeanne and Nancy for at least several decades, and I think she still has some as souvenirs.

We of course had heard of the famed beaches of the Côte d'Azur and probably were anticipating counterparts of those beaches we had experienced in Miami or the Gulf of Mexico, but these famed beaches in France were mostly <u>stones</u> – at least their upper reaches were. People reached sand and water by walking down narrow board walkways over upper sections of <u>"beach" paved completely with stones</u>. Even this was OK, except that the walkways were narrow enough to give confirmed worriers like me some concern about accidentally brushing up against the hordes of bare chested nubile women who populated them, and the awful legal problems that I might face as a consequence of the touching. The views were superb (mostly) but the possibilities of accidental contact were real in such close quarters! The Mediterranean waters were all that I expected, but <u>getting to them</u> was "disconcerting" for an unreformed New Englander! That was my final opinion!

We liked every part of that last trip to Nice and Monaco. The weather was ideal as was the lodging. The hotel was small and friendly, and also on a headland directly overlooking the city of Nice and the Côte d'Azur beyond, so the view was absolutely spectacular, and yet we were only minutes from downtown. The city was tourist-oriented to some extent but within easy walking distance of the hotel – close to an ideal arrangement for us.

Taken in its entirety, France suited us just fine in every way, and it remains one of our favorite European countries.

Figure 14. Members of the ICES Working Group on Introductions and Transfers of Marine Species at a meeting in France in 1986.

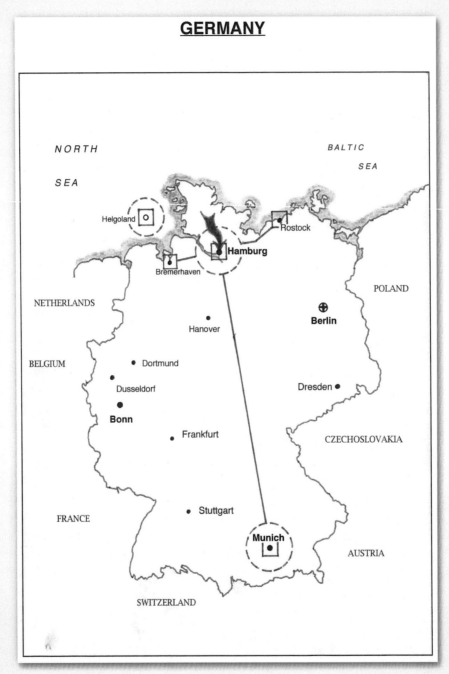

Map 13. Germany.

CHAPTER 16
TRAVELS IN GERMANY

In developing this section about personal travels in Western Europe, I was somewhat disturbed about how little Joan and I had actually traveled in Germany, the largest country in Western Europe and the country from which my father had come. Fortunately, we made several extended trips to and beyond the Hamburg area, which helped to make up to some small degree for the obvious geographic distributional defect.

Of course, the highlight of one of our visits was a 30 mile sea voyage to the island of Helgoland and the marine research institute there, but I was impressed with the reconstruction of the city of Hamburg itself, which had suffered from extensive bombing in World War II. Hamburg is today a graceful modern industrial city, which, coincidentally, houses a well-known university and several well-known marine research laboratories that hosted our presence there on repeated occasions, and sponsored extensive explorations of the surrounding country-side.

I am sure that the narrowness of our exposure to Germany as a whole was because much of the ocean-oriented research and development of the country is centered in the Hamburg area, so it is entirely logical that much of my official travel in the country would be focused there, to the virtual exclusion in this tome of much of a large and beautiful country that was therefore not dominant on our travel agenda. I will say, though, in partial attempt at some justification, that some of the directors of laboratories in this region were among the best that I have encountered anywhere. Their approaches to the position of scientific research laboratory director were exceptional and sometimes unique; they were often role models for many of us in comparable positions in other countries.

As an example of the hospitality offered, a good friend and senior scientist at the Helgoland Laboratory drove me, at unbelievable speeds, in a new Mercedes, down the Autobahn from Hamburg to Munich, and then on to Venice in Italy to attend the World Conference on Mariculture in 1981. I saw as much of Germany that I was to see on that high speed trip. We crossed most of the breadth of Germany in that one day of travel at tremendous speeds, stopping for the night at a hotel outside Munich run by the United States Army. (It just so happened that the research center that I directed at the time, the Middle Atlantic Coastal Fisheries Research Center, was located on the grounds of Fort Hancock, which was also a U.S. military base for missiles protecting New York City. I found out at the military hotel desk that a Research Center Director held the military equivalent rank for housing of lieutenant colonel, and my German friend, who was driving, was rated as a staff sergeant). We spent the night with the military, and then went on through Austria and part of Italy to the Venice meeting, after a good American breakfast with the U.S. Army of Occupation.

Hamburg was and is the center for much of Germany's oceanographic research and development, so it was logical for our scientific interests to center there, to the virtual exclusion of the rest of a large and beautiful country. This has been a distinct handicap in writing a chapter titled "Germany," but it is a very logical reason for the absence of descriptions of much of the large and prosperous country seen only at high speeds from an autobahn or from the air, or not seen at all because of lack of proximity to an ocean. I, for example, have always wanted to spend some leisure time in Bavaria, since my forced travels there late in World War II. My recent total thus far has been limited to an overnight stop at an army operated hotel off the autobahn near Munich. However, if what we have seen and enjoyed of Germany in the Hamburg area is any indicator of the whole nation, then this is definitely a country that should be high on any traveler's list – scientist or otherwise. I recommend, purely on a trial basis, a boat trip on the

Rhine or a visit to the restored city of Berlin, or any activity that involves local or long distance travel by any conveyance other than air.

Unlike so many other veterans of World War II, I did not go back to the narrow part of western Germany that our infantry division had occupied near the end of World War II, and strangely, I had no desire to go there again.

Map 14. Low Countries: The Netherlands, Belgium And Luxembourg.

CHAPTER 17

TRAVELS IN THE LOW COUNTRIES: THE NETHERLANDS, BELGIUM AND LUXEMBOURG

Very early in my career as a research biologist for the federal government, at the Biological Research Laboratory in Boothbay Harbor, Maine, we were joined for a summer by a young female professional staff member from the Shellfish Research Laboratory at IJmuiden in the Netherlands. She was studying, at the time, blood groups in eels – a close counterpart to some of my work with sea herring blood groups, and a logical new area for collaborative research in fish population studies at that time.

The summer passed productively and very pleasantly, and was the beginning of a career-long friendship for both of us, based on occasional appearances at international meetings of joint interest. Joan and I visited her laboratory at IJmuiden, which also had a world-wide reputation in molluscan shellfish research, and was located close to Amsterdam, one of our favorite European cities.

We made subsequent trips to the Netherlands – to Amsterdam of course, but also to The Hague and to the coastal town of IJmuiden. Our introduction to the country was prompted, as I recall it, by a large international scientific meeting, held in downtown Amsterdam near the Museum of Natural History. It was at this early meeting, when we began talking with other participants, that I began to realize that there existed a community of very good scientists from many countries (but mostly European) who knew one another and who seemed to be delighted to be in the presence of one another. For these scientists, technical meetings were important to keep abreast of new developments in science, but also as a meeting place for friends and colleagues.

This, to me, was an extremely important observation – and one that changed my entire perspective on technical meetings. Such meetings should have many purposes. The most significant must remain the transmission of new scientific information to peers and colleagues. But beyond this is the creation of the whole informal infrastructure of science – where peers become friends and not competitors, and where technical meetings are to be enjoyed as well as profited from. This insight has guided my entire approach to scientific meeting travel ever since.

My visits for immunological conversations with my Dutch colleague brought me in occasional contact with her laboratory director, Dr. Pieter Korringa, an outstanding shellfish scientist and administrator. He unconsciously became a career model for me in many ways, and I was constantly overwhelmed by the breadth of his achievements as director, and his skills at so many of them. (He, for example was smoothly multilingual, had published several books in his specialty, and was a superb planner and implementer of international scientific meetings – all from the base of a small shellfish research laboratory on the outer coast of a small European country!)

I, for many years thereafter, was able to watch this exceptional laboratory director function at ICES committee meetings (and occasionally at the higher administrative levels of that organization). Each performance was a lesson for me, even though I am typically a slow learner. I know that I profited significantly from even fleeting contacts with this admirable man.

Of course at least some of our visits to Amsterdam had statistically to occur in spring time, when tulips are still an annual frenzy in The Netherlands. While I was in technical meetings of various kinds, Joan was free to go where the action was and where blooms were at their peak. (Some of the meetings added spouses' tours as parts of their programs.) She went by train or bus with other spouses to vast fields of tulips in every stage of development or destruction – or in rare moments of glory, in every rainbow color seen in only a few places on earth (we have a small counterpart in the state of Washington). She looked and placed orders and brought back samples to the hotel room. It was a special time and place for her, as it would be for any flower lover, and an experience that is still vivid in her mind, so many years later, and one that she was able to repeat a few years after that.

It was on one of these early trips to The Netherlands that Joan and I decided to rent a car and visit parts of Belgium and Luxembourg that I had "passed through" much earlier, during World War II, as a medic attached to an infantry reconnaissance platoon during the Battle of the Bulge. Most of the towns in Belgium that had been severely damaged during the war had, by the time of our visit, been repaired and rebuilt, although here and there an isolated ruin of a house could be seen half-hidden by foliage. About the only change that I could see beyond this repair in Belgium that had been affected directly by the war was the almost universal appearance near the town centers, of war memorials, often consisting of tanks or cannons or armored vehicles on cement platforms. Probably what interested me most was that some of these war memorials were of <u>German</u> rather than <u>Allied</u> origin – here and there were <u>German Panzer Tanks</u> or <u>German 88 millimeter cannons</u> as war memorials, and not always Allied equipment. All these items may have rusted away or have been replaced during the intervening years until this writing, and I have never investigated the issue further, but I have often wondered about this observation, and the basis for its reality in so-called "liberated" countries, and more broadly on the "transience of glory."

On a more topical positive and relevant level, in ICES interactions in working groups and in Committee meetings, I have always found the Belgian scientists to be informed and responsive colleagues, and full participants in the deliberations of the working groups in which they participated.

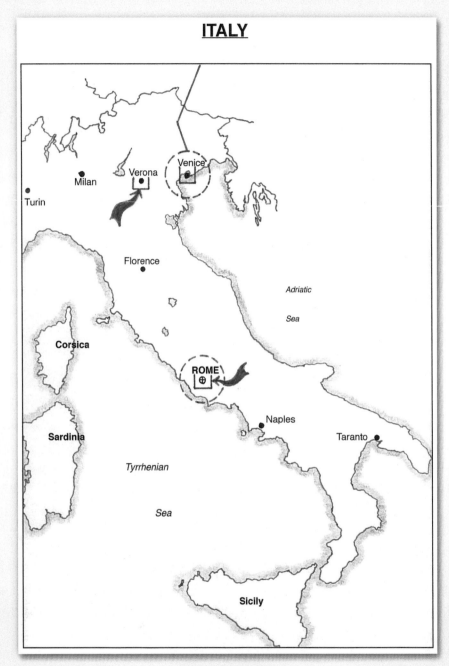

Map 15. Italy.

CHAPTER 18
TRAVELS IN ITALY

Italy is geographically not immense, but large events take place there, and they have done so for us. Four events stand out in our memories of visits to that country for vastly different reasons:

A. Joan's extended visit to the papal facilities in Rome.

B. My impromptu but memorable nighttime walking tour of Rome.

C. The 1981 World Aquaculture Society meeting in Venice, and

D. A brief but memorable visit to Verona, one of Shakespeare's favorite cities in Italy.

(A.)   THE VATICAN IN ROME

A visit to the headquarters of the Roman Catholic Church was of course de rigueur even to fallen-away Catholics like Joan and me, but our visit to the papal entity in Rome was a major project to her because of her family's extensive Catholic religious history. She was able to spend most of an entire day touring the various chapels and rooms and buildings of the Vatican – something every good Catholic hopes for – while I, of course was in some kind of technical meeting in another part of the city. (We did get together that evening for dinner with FAO-UN colleagues, though, as I recall it).

(B.)   ROME AT NIGHT

While visiting Rome alone for a now long-forgotten FAO-UN consultation on West Africa I had the superb good fortune of making the acquaintance of an expatriate male American serving a tour of duty in Rome, who had long ago fallen in love deeply with the city, and who tried in one unstructured evening's walk to explain to me and to show me why the city was so important and so special a place. We walked literally for miles through the late evening quiet to places like the Coliseum, the Vatican area, even some of the business as well as historic sections, while he kept up a running discourse that melded history with the present day, and in some instances emphasized the great historical significance of the buildings and ancient Roman structures that we were passing. His descriptions of the present, when combined with history, made the evening a priceless one for me, never to be forgotten – the kind of relaxed informative treatment that every visitor to an historic place deserves but rarely if ever receives.

We parted very late that evening, and I never saw him again, if only to thank him one more time for a wonderful evening of rare insights about that historic city – the most memorable that I have ever experienced. (*Arrivederci* Roger, wherever you are, and thank you again! It was a night that I will never forget)!

(C.)   INTERNATIONAL CONFERENCE ON AQUACULTURE HELD IN VENICE, ITALY IN 1981*

I know that I have written about my absolutely most favorite scientific society, the World Mariculture Society, before in this book, but its meetings have been so varied and so spectacular that they deserve more attention here and elsewhere too. This is a truly international organization, that holds its annual meetings in widely dispersed geographic locations around the world and has an international membership. The best meeting of that society, in my opinion, was held in Venice, Italy, in 1981, in the tourist heart of the city. Outstanding meeting arrangements were made by European members of the Board of Directors of the Society, so all the sessions were held in the ancient classical center of the city, as were all the hotel assignments. The program was stretched to allow much free time for sightseeing and gondola riding, and the technical sessions were as excellent as the after-hour activities.

It was, in my estimation, the kind of scientific meeting that exists mostly in fantasy, but occasionally becomes a reality – and this was an example. It was also held at the termination of the year that I served as President of the Society – a reality that provided me with no end of personal satisfaction, given my propensity for participating in world scientific events that matter.

--

Earlier in this book I have mentioned my most favorite scientific society, the World Mariculture Society, which sponsored the Venice Aquaculture Conference in 1981. That society changed its name and broadened its perspectives by becoming the "World Aquaculture Society" in 1986!

--

It was, of course the scientific meeting that I will always remember, not for the technical content, but for all the human interactions that were involved in planning and conducting such a successful venture, especially the impeccably perfect choice of the meeting milieu in Venice by the European members of the Board of Directors of the Society! The setting was superb and the sessions equaled them – a perfect combination for a scientific meeting, and typical of those conducted by the World Mariculture Society.

(D.)   TRYING TO ACT LIKE A GENTLEMAN FROM VERONA

I had heard vaguely, from uncertain sources, that there existed in the city of Verona in the extreme northern-most part of Italy a Roman amphitheater that was still reasonably intact and in use (but not for its original purpose). I also recalled, even more dimly, that several of Shakespeare's plays were based on characters from the city of Verona. Therefore, when an announcement appeared in the late 1980's about a major international symposium that was to be held in Italy, in the city of <u>Verona,</u>* on the subject of the future of marine aquaculture (a subject close to my research interests at the time) I was interested. I became more interested when the symposium organizers offered me a liberal all expenses paid trip if I would present a paper on my specialty: "The Role of Disease in Marine Fish and Shellfish Aquaculture." Being something of a Romantic (sic), I saw myself delivering my scientific speech from the podium of the Roman arena – a first for me (and no doubt for most other scientific speakers)! But – sensibly – such was not to be! The location of the symposium was <u>not</u> the arena but a modern <u>commercial building downtown</u>. All those centuries of history in that city, including recent hostile events between Austria and Italy before and during World War II, and so little time to explore! (I did get a glimpse of the amphitheater, though, on the way back to the airport, and I have often been truly amazed and dismayed by my sheer lack of planning to exploit this venture in foreign science! A good lesson in how <u>not</u> to visit a foreign city bursting with historical significance! The lesson: "Be smarter than I was; enjoy the ambiance of your surroundings! And most of all, for a visit like this one to a foreign city with multiple historic features, <u>always</u> bring your wife and family for a true encounter with history!"

These small vignettes are personally important to Joan and me, but they are totally inadequate to cover the totality of Italy. Our several visits to that country have been pleasant in some ways, but very inadequate to develop real understanding about it and its people. This may be a normal tourist's reaction to brief visits, but we certainly enjoyed the segments that I have discussed briefly in this chapter, especially, I think, those that centered on Rome itself.

--

Verona was the venue for three of Shakespeare's plays including "Two Gentlemen From Verona," and is the site of the third largest (and still utilized) ancient Roman (A.D. 30) Amphitheater, with a seating capacity of 25,000!

--

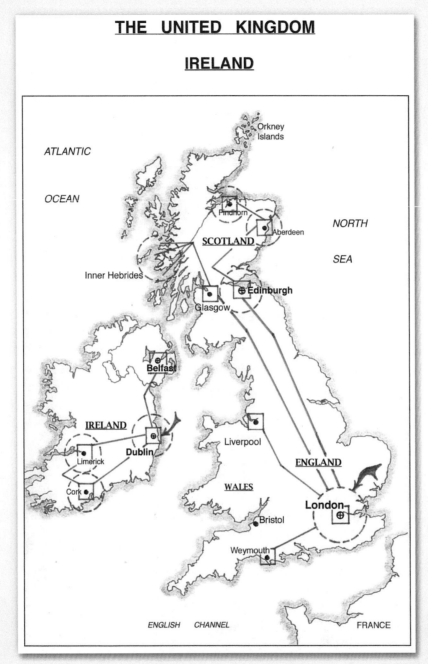

Map 16. Britain (United Kingdom).

CHAPTER 19
TRAVELS IN BRITAIN (THE UNITED KINGDOM)

Of all the interesting parts of the United Kingdom (Britain), we visited London a number of times, Scotland for several extended periods, Wales only briefly for an ICES meeting, and the southern coastal town of Weymouth, where a major marine laboratory was located.

Most if not all of these visits were the consequences of scientific meetings of many kinds beginning (as examples) with an International Symposium on Oil Pollution of the Seas, held by the Royal Society of London, and a week-long meeting of the ICES Working Group on Introduced Species held in Aberdeen, Scotland. At another time there was an invited paper for the International Society of Parasitologists, meeting in Liverpool.

I am sure that our forays among English-speaking foreigners – and those from the U.K. itself, were definitely more relaxed and productive because of the absence of a language barrier, although I was constantly surprised by the extent to which English was understood by members of the scientific community in general.

Figure 15. The author in Piccadilly Square, London, 1970.

One of our tours in Scotland took us to its extreme northernmost coast facing the Moray Bay to the tiny town of Findhorn. My son Jim and his wife were there at the time with the Findhorn Foundation, a religious group, at the moment restoring what seemed to be an ancient castle belonging to the Foundation. We stayed in the only hotel in town for a few days, touring some of that wild but beautiful north coast of Scotland before moving back south to Aberdeen and other interesting places in Scotland.

While in the north of Scotland, outside the town of Findhorn, we came upon a famous relic of the ancient sailing ship days – the remains of a Martello Tower – a stone tower used by townspeople in defense against marauding ship's crews. The presence of this dark and silent place of ruin in the forests at the edge of the coast, intrigued me enough to make a small study of it and other counterpart small circular stone ruins scattered along the coastlines of northern Europe. The summary of that study is appended here.

THE MARTELLO TOWER OF FINDHORN, SCOTLAND

Symbols of early Europe, now fading into the forests and dunes, on the outskirts of coastal towns, are remnants of circular masonry towers built for protection of townspeople against marauders from the sea.

The basic circular design was adopted by the British during the Napoleonic Wars of 1805 and 1808 as coastal defensive forts called "Martello Towers". The name has acquired general application since then for any small circular masonry defensive structure of advanced age in a coastal site, far beyond the basic protective concept of the original British design. These so-called "Martello Towers" and their predecessors, have had a strange fascination for me, and undoubtedly for others: first, because their remains exist at all, and they are early evidence of people banding together locally against common enemies from the sea, but also because they are reminders of a dark and dangerous period in human history that is, unfortunately, part of our heritage.

We personally encountered the remains of coastal defensive towers on the extreme north coast of Scotland and one of similar circular design on the north coast of France. The one on Scotland's north coast, a true Martello, was a local landmark and tourist attraction; the circular masonry tower on the French coast was in advanced disrepair, and was considered to be a dark and mystical place from early history, equipped with magical powers.

We spent much of our time in Scotland in and around Aberdeen, mainly because that was the location of the ICES Working Group Meeting. We enjoyed the city, especially its many associations with the sea. My predominant impression of Aberdeen, Scotland was one of friendly people living in a city built entirely of gray stones. We, Joan and I, had occasion to visit the city for an entire week during sessions there of the ICES Working Group on Introductions of Marine Animals. As it worked out, a prominent and active member of the Working Group lived in Aberdeen in a big ancient stone house that he was restoring and he had the restoration far enough along to invite the entire Working Group to several evenings of entertainment and relaxation. So the meeting was a great success, scientifically and socially. (We learned later that he had completed the restoration, and was finally able to enjoy his long-term project).

CHAPTER 20
TRAVELS IN IRELAND

Today I want to write about some impressions of Ireland, where Joan and I (and the members of the ICES Working Group on Introductions of Non-Indigenous Marine Species) spent an extremely pleasant week in Dublin, the capital city, followed by a week or more of visits to some favorite tourist locations such as the Ring of Kerry, Cork, Limerick and Killarney, completely on our own. We were partially prepared for the Irish good feeling and openness, since Joan had grown up in a Massachusetts neighborhood with a substantial sprinkling of second and third generation Irish immigrant families, and had known and liked most of them.

I, typically, had prepared myself for a darker side of the Irish nature by reading some of the literature on the awful starvation period of the late 1800's with such books as "The Great Hunger" and "The Famine Ships," – but, as usual, my dark preconceptions were totally wrong!

I had read often of the British landlords' propensity, during that awful period of starvation and want – the potato famine – for destroying homes before their occupants had even vacated them, so I was doubly impressed with the many remnants of entire villages – streets of stone houses without roofs or interiors – towns without a single occupant – with remnants still standing in many locations, on the coast and inland as well. I can still get the feeling of major human catastrophe here, now represented inadequately only by stark decaying walls of what were at one time vibrant human habitations – a sight that I will never forget.

But the Irish have recovered from catastrophe, with vigor and self-respect, that to me is a good trait in a worthwhile subgroup of the human species! The people of Ireland that I met were outgoing, pleasant, upbeat, and genuine! Their scientists too were fully capable and in some cases leaders in their fields of specialization, but in all cases a pleasure to work with. (I remember with great clarity that the presence of an Irish scientist at an ICES working group meeting was always considered as a positive factor contributing to the pleasure and success of the sessions). But I think it was the people themselves, in all the towns and villages of Ireland that we visited as tourists, that made the best impression, with any kind of contact with us. It was, simply put, a genuine pleasure to be among them, whether they were shop-keepers, cab drivers, people on the street, policemen or fellow scientists. They uniformly projected a positive, upbeat, optimistic, somewhat whimsical outlook on life, living, and science, with no traces of their awful recent history of starvation and forced removal at the hands of the British in the late nineteenth century. This kind of signature up-beat positive attitude seemed to prevail among the Irish people wherever we went, but was especially notable in their homeland as well as abroad.

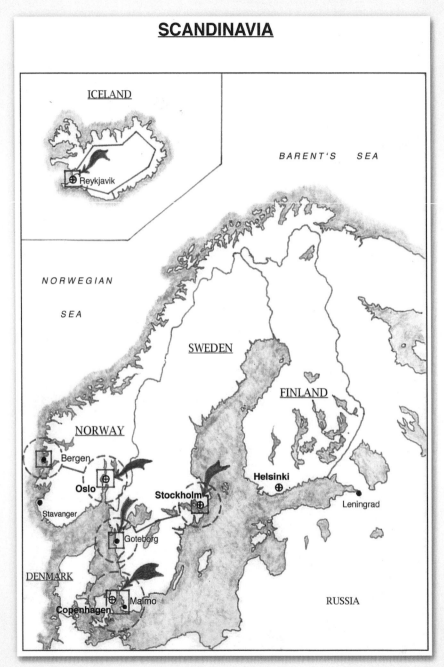

Map 17. Scandinavia.

CHAPTER 21

TRAVELS IN THE SCANDINAVIAN COUNTRIES: DENMARK, SWEDEN AND NORWAY

Some of the most beautiful cities of the world can be found clustered in Scandinavia, beginning, of course, with Copenhagen in Denmark and going on to Stockholm in Sweden, and Oslo in Norway. We have enjoyed repeated extended official visits to each of them – not only because of the physical beauty and grace of each place, but for the European friends that we made in each one. We were free to roam through these historic places in off duty hours in congenial groups that often included natives of the country we were in – what a pleasure it was for all of us!

In retrospect, our experiences in the Scandinavian countries closely rivaled those in France or Ireland or the U.K. for sheer pleasure as tourists and as memorable places to return to in that "someday" that rarely comes. Also in retrospect, it was principally the "people interactions" that provided most of the pleasures (although the scenery at times was overwhelming in its own right in each country).

Because of the nature of my profession, we tend to offer a "waterfront" view of many of the countries that we have visited. I want it to be known, though, that we have occasionally invested in a rental car for an inland tour of many of the countries in Europe. This we have done repeatedly in the Scandinavian countries as well as the Low Countries and occasionally elsewhere, if the location seemed to call for inland travel. In the case of Scandinavia, however, so much of interest is coastal that inland forays seemed unnecessary.

We shouldn't be too obvious in our personal likes and dislikes of people on a national basis, beyond the admitted tendency toward English-speakers – but if pushed a little, it would be the Scandinavian scientists – from Norway, Denmark, and Sweden, who would be our next choices for compatible international partners in scientific affairs. This might well be a consequence of repeated visits to those countries because of their joint membership with the United States in the International Council for the Exploration of the Sea (ICES), a venerable European scientific organization with close governmental ties and a multiplicity of meetings, commissions, standing committees, and a history of profound influence on ocean affairs. It is an organization in which the northwestern European countries invest their scientific capital heavily, in terms of laboratories, vessels, research funding and scientific personnel, as well as in statutory meetings and working group attendance at joint meetings.

Figure 16. Our favorite meeting place in downtown Copenhagen: The Radhusplatz!

The beautiful part of this picture is that the scientists that I met, who represented these Scandinavian countries – the laboratory directors and senior scientists – are usually what would be described as "world-class" – erudite, productive, thinking people who are pleasurable to be with on any occasion, scientific or otherwise, and all are multi-lingual. One trait that I really came to appreciate (in my role as working group chairman) was their "in your face" frankness; they always gave an honest opinion, whether favorable or not, on any subject -- and the rest of us in any technical or advisory group appreciated this trait immensely.

Among foreign scientists I have met, I have noted that Scandinavian scientists can be severe critics but also significant supporters of American research efforts, and I, personally, have always enjoyed being in their company! Taken as a group, they are wonderfully intelligent, communicative, and active participants in any complex science-related discussion.

Figure 17. A view of the modern downtown waterfront in Copenhagen, Denmark.

Map 18. Eastern Europe.

CHAPTER 22

TRAVELS IN EASTERN EUROPE:
POLAND AND RUSSIA (SOVIET UNION)

Our visits to Eastern European countries were limited to Poland and Russia (then part of the Soviet Union), and were comparatively brief when compared with those extensive visits to the Western European countries. Furthermore, the visits preceded major changes in the governance of both countries: the overthrow of communism in Poland and the dissolution of the Soviet Union. We were there, early-on, during those exciting times, but of course played no role, except as interested bystanders to world events as they were happening and where they were happening (that is not quite true, since we almost accidently and innocently got involved very peripherally in the diplomatic maneuverings preceding the fall of communism in Poland, as you will read in the following pages).

Let it be known early in this discourse that our brief experiences in Eastern Europe provide us with no reasons to encourage tourist travel by anyone to that part of the planet. Both the Polish and the Russian people seem to be suffering from their admittedly awful experiences during and after World War II. Whereas Western Europe seems to have assumed a positive perspective on existence, Eastern Europe seems somehow still trapped in the grayness and negativity of the horrendous past – and probably with good reason. Here and there, especially in Poland, are brief flashes of good cheer and even pleasure, but they are rare and artificial and not worth the search.

So, my suggestion is, go to Eastern Europe if official or organizational duties require your presence, but otherwise seek brighter, happier places!*

--

    I have some miniscule vested interest in the political situation in Eastern Europe since <u>I was there</u> when the lines were first drawn that soon after that would become the "Iron Curtain." At the end of World War II, I was with a 26th Infantry Division Reconnaissance Platoon stationed in a small town of Oberplan in Czechoslovakia. One day we received immediate orders to withdraw to Steyr in Austria because the zone that we were in had been ceded to the Russians in negotiations. This, though I could not recognize it at the time, was the first "drawing" of the so-called "<u>Iron Curtain</u>" that has played such a major negative role in the future development of the Eastern European countries.

--

Figure 18. Red Square in downtown Moscow.

You may find, in some Polish and Russian individuals, a reawakening of some earlier spirit of freedom and even occasional gaiety, but it will be rare. Go if you must to these places, but preferably not as tourists, unless there has been some substantial change in general attitudes and perspectives in more recent times.

The world literature continues to accuse Eastern Europe (including all of Germany) of "delaying and even reversing human progress away from its "animal past" before, during, and after World War II. This literature seems to be having a new peak just when the last of the perpetrators are disappearing from the planet due to natural causes. As history is showing, recovery from those dark days has been slow and sporadic throughout Eastern Europe where the terror and the killing was most severe.

So – go to Eastern Europe if organizational or institutional duties require your presence, but otherwise seek cleaner brighter places elsewhere on the earth. (That would be my rather dismal recommendation.)

CHAPTER 23

TRAVELS IN POLAND

Despite my decidedly negative introduction to this section on Eastern Europe, I find that the visits we made to Poland were on the whole reasonably pleasant ones. Most of them were to Warsaw, but we did get up to the Baltic Coast to Gdansk and Gdynia on several occasions. The time spent in Warsaw was the best, though, at ICES meetings of various kinds, where we gathered several times with scientists from all over Europe for week long meetings on various topics of concern at the moment.

I remember very well and with great pleasure, walking the streets of Warsaw, including those of the reconstructed "old town," with congenial members of ICES from many different countries of Europe, and thinking that, as a scientist, there was no place in the world that I would rather be in and no group of people that I would rather be with at that particular moment in time! (An important kind of feeling in a foreign city, and one that I remember so very clearly!)

It seemed that whenever we were in Warsaw it was raining or at least threatening to do so, so our impressions may have been a little unfair. We were in the city for ICES meetings or for joint meetings with Poland on bilateral fisheries research projects, or as a departure point after visits to other parts of Poland. Warsaw was of course a city of much World War II history, principally one of death and destruction of the Jewish community there, with abundant memorials of that infamous period in Polish history. But the city itself has revived remarkably, with new hotels and a bright downtown section, a completely rebuilt old town section, and suitable memorials of the dark period of its existence.

We didn't get to Poland quite as often as we might have as part of my role as an ICES working group chairman. One trip to Warsaw had to be cancelled because of "visa problems"* but otherwise our visits to that country were pleasant and instructive.*

--

I have a strong but completely unsubstantiated suspicion that some of my problems with the Polish Embassy were self-generated, because of my own stupidity. I have been accused on rare occasions of telling so-called "laugh-lines" that are a "little out of line" (read "<u>much</u> out of line"). These tend to contain minor ethnic jests, like the "Newfie" jokes told by Canadians to belittle Newfoundlanders. My stupid error was telling a Polish story once too often to the wrong audience (and I am not of Polish extraction) about Warsaw as a tourist destination.

The story – a laugh line, and a dumb one, told too often to the wrong audiences, was about a contest with a first prize of one week all expenses paid trip to Warsaw and a second prize, an all-expenses paid trip of two weeks, to Warsaw! A stupid, stupid story for adult conversation. (I can't believe my stupidity!!)

But there is more about this for which I have only circumstantial evidence. Soon after my performance, my passport visa for Poland was inexplicably delayed by the Polish foreign office in Washington so that I could not and did not attend an ICES Working Group Meeting in Warsaw (as ICES Working Group Chairman).

I took this as a very real and very hard lesson in correct international manners for amateurs and idiots, and one that I will never forget, even if the coincidence may have been in my mind and not real at all. (And I found in later visits that Warsaw was a lovely city for which I have many pleasant memories!)

--

Our most recent tale of experiences in Poland took place on the main highway from Warsaw to the Baltic Sea in 1989. Dr. and Mrs. C. J. (Magoo) Sindermann are tootling northward in a U.S. Embassy station wagon headed for a legitimate inspection trip of a contract plankton sorting unit, located near the city of Gdynia on the Baltic Coast, and funded by U. S. dollars.

They (Dr. and Mrs.) were on a mission, regardless of the political unrest around them – to inspect the plankton sorting station funded by United States (NOAA) dollars. They were also in a particularly volatile part of the country during the decline and fall of communism there (the northern coastal (Gdansk-Gdynia area)) – an area from which the U. S. Embassy in Warsaw felt that it needed current information about offshore Soviet Union blockade activity.

The inspection of the plankton sorting facility was carried out <u>pro forma</u> and we were free to enjoy the delights of the city of Gdynia. Joan saw and joined a long line of locals and at the end received one roll of incredibly harsh pink toilet paper! We then retreated to our hotel for the afternoon and evening.

The embassy staff people who had accompanied us reappeared the next morning and we began the long journey south to Warsaw (gas was carried for the whole trip in five gallon cans behind us in the rear of the embassy station wagon). Except for the absence of formal pit stops (gas stations) along the way, the trip, taken during a tense political period was instructive technically, and pleasant. We received the tiniest taste of how our country functions in foreign countries, and we liked it!

We did have just the faintest feeling that we were being used as decoys by the embassy, but the scientific objective of the trip was routine and entirely legitimate from my perspective. The embassy simply facilitated what would have been a difficult technical inspection trip, considering the instability of the period, and we appreciated their help, while recognizing the advantages to both parties in the trip to the Baltic during such an exceptional time in world history!

In concluding this chapter, I would point out that any discussion of Poland as a destination for informed travelers must include its at times simply awful history during and long after World War II. Even now, more than half a century after those dreadful days of mass deportations and murder, violent racial upheavals, foreign domination, extreme poverty, and population dislocations, the Polish people seem to be struggling for some sort of normalcy in everyday life.

I believe strongly that any visitor to Poland, or for that matter to any Eastern European country, must recognize the possibility of an existence of that kind of history, which is only slowly dwindling away, as are those who were direct participants in the atrocities against humanity that were perpetrated in Poland and elsewhere in Eastern Europe in particular.

So, go there if you must, in good spirits, with a positive attitude, and try to avoid discussions with local scientists, or other people, of those very dark days of Polish history!

CHAPTER 24
TRAVELS IN THE SOVIET UNION (RUSSIA)

We were in the Soviet Union – mostly in Moscow and Leningrad – during an interesting period just before the dissolution of the Soviet empire in 1991. Our visits to those cities occurred during a period of temporary receptivity by those in power to joint discussions on scientific matters, particularly those related to fisheries. Our very elaborate 10 day visit to Leningrad (now St. Petersburg) was sponsored by the Soviet Academy of Sciences – to the extent that all expenses were paid and a tour bus was made available to Joan and other "accompanying persons" for the full week of the formal meetings. Their "accompanied tour" covered every historic site within a 20 mile radius of St. Petersburg, and was a trip that she has always remembered. Each morning a tour bus with a guide appeared at the hotel to pick up Joan for a full day (including lunch) of sightseeing in and near one of Russia's most historic cities. Places like the winter palace within and outside the city were included, and requests were granted for locations not on the prescribed list of destinations.

Meanwhile, during lunch breaks in the Soviet Academy of Sciences program, I did get a few minutes to see some of the statues and former palaces of the Czars that the city is noted for. For me, though, the crowning Leningrad experience was a personally narrated weekend guided tour by the Academy director who took Joan and me through some of the now grass-covered battlefields of that amazing period of World War II when the citizens of the city of Leningrad withstood a German siege for three awful years – and he was there all during that time of starvation and death! A gift from him to us beyond any price!

Slowly, for us, the meaning of the term "official guest" in a communist country became clear, and usually enjoyable for all its "perks" – however transient they may have been. We did meet good people and good scientists, regardless of the venue, and the whole experience at Leningrad was a positive one for both of us – Joan for the carefully scripted constant care of what must have been security employees on tour buses and in historical points of interest – and for me the brief but cordial association with Soviet counterparts in my fields of technical interest. It was a thoroughly pleasant week, capped by a personally-led tour of historic World War II sites by one of the prominent figures in my area of science.

Despite the kindnesses shown to us by the Soviet Academy of Sciences in Leningrad, we faced the harsher realities of Russian existence during several subsequent visits to Moscow to discuss Atlantic fisheries research. Here we encountered the impersonal but thorough hands of what I believe was called at the time the KGB or some acronym. We usually stayed at the huge Hotel Rossia,* just below Red Square, and an admirable refuge once you put yourself completely in the hands of your giant female "floor captain," who seemed to be in charge of everything, once you had surrendered your passport.

We rarely, if ever, felt really comfortable in the Soviet Union, even though individual Russian scientists did their best to make our visits pleasant ones, insofar as they were able to. I guess that one way to describe our attitude was that we appreciated exposure to the places and to other professionals but we might not choose to go back there as tourists.

--

Joan has always been a great travel companion, but she occasionally gets into what I call "situations" when I am not present – like the cross-eyed taxi driver in London, or standing in line for an hour in Gydinia, Poland for one roll of pink toilet paper (we were staying in an elite hotel at the time). But her most memorable "situation", and one we still laugh about, occurred in Moscow: We were staying at the giant Hotel Rossia, just below "Red Square". We enjoyed the comfort of having a very large female employee as "floor captain", ready to solve any problems from a desk on "her" floor (but, unfortunately, she spoke no English at all). We weren't enthused by the security police at every exit door, but they quickly became part of every day life.

One unusual incident occurred, though, during our stay at the Rossia, that Joan and I have told repeatedly: I was in a ground floor meeting room (a bar) for late-evening discussions with colleagues, when a great commotion arose at the elevator gates. It was eleven p.m. and the whole elevator system closed down precisely at that time (a fact unknown to us). A woman had been trapped by the closure on an upper floor, and was screaming for help, but no-one seemed to be available to restart the elevator system. (The technicians had presumably left the premises immediately after pushing the "stop" button.)

My first act, on hearing the cause of the commotion, was to offer a five ruble bet to members of my discussion group that the woman trapped in the elevator was my wife Joan.

After enough frantic activity, someone found the key to restart the system – enough to get the trapped passenger to the first floor. The elevator doors opened and a woman stepped out, cool and smiling, and waving to the cheering crowd in the hotel lobby – it was Joan, of course!

--

Our impressions of Russia (or the then Soviet Union) would have to be described as "mixed." On the one hand we were treated with the utmost courtesy when our identities as official guests were known, but on the other hand we had to observe the rough impersonal ways in which their own citizens were treated all the time (unless they were officials of some sort). Based on this dreadfully inadequate sampling we would have to describe independent tourism in that part of the world as "challenging," and even "to be avoided"!

Both Joan and I appreciated our visits to the Soviet Union – she for the total immersion in Russian history and I for interactions with excellent scientists. Our week in Leningrad (St. Petersburg) was exceptional in many ways and our visits to Moscow very informative, insofar as time permitted.

PHASE V
GENERAL CONCLUSIONS ABOUT SCIENTIFIC TRAVEL IN FOREIGN COUNTRIES
(1991 – PRESENT)

PHASE V
GENERAL CONCLUSIONS ABOUT SCIENTIFIC TRAVEL IN FOREIGN COUNTRIES
1991 - PRESENT

Late in my tenure as Center Director of the Middle Atlantic Coastal Fisheries Research Center I applied for and finally received what was called an "Intergovernmental Personnel Act Appointment," which allowed me two years to complete my government career with involvement in research and writing of my choice at government or government subsidized facilities of my choice nationwide. I elected to use the program funds for research and writing periods at a number of institutions, including the Rosenstiel School of Marine and Atmospheric Sciences at the University of Miami, the Oxford Biological Laboratory, and the Northeast Fisheries Science Center at Woods Hole, MA. I was able to complete my major two-volume 1000 page book on fish and shellfish diseases, "Principal Diseases of Marine Fish and Shellfish," and several other books, technical or otherwise, with funding assistance from this far-reaching and extraordinarily liberal federal act.

That funding has of course long since gone, but the research and writing habits developed during the appointment period have remained with me to the present day, and are largely responsible for this Phase V of the book: contemplation of what has gone before, and a summarization of at least some of the insights gained!

I want to use this final Phase – Phase V – of the book to spend some time on generalizations and summarizations about travel for scientific purposes. Originally, I envisioned a chapter covering all the elements of this phase, but it has grown beyond my control to five extensive summary chapters:

Chapter 25. The Pleasure of Foreign Travel as a Scientist;

Chapter 26. The best of all travel destinations: The United States;

Chapter 27. Leadership roles for scientific research laboratory directors in international science;

Chapter 28. Achieving successful international scientific relationships; and

Chapter 29. General conclusions about scientific travel in foreign countries.

I have tried, in these concluding chapters of the book, to consider the value, significance, and even the pleasures of all this international travel, from a scientific as well as a personal perspective. My findings are, like my travel agenda, severely limited by lack of time for adequate exploration of some of the remarkable places we have visited, but at least we were there – and in some places like Taiwan, Poland, and Russia, we were there during momentous times in history.

One final personal point to be made here is that <u>scientists (or other professionals) whose jobs require trips to foreign countries should go to those places regardless of costs, with a spouse </u>and <u>more or less occasionally with other family members!</u> I am not pleased with my own response to this dictum! In retrospect, we should have done much much more, especially with a growing family, as ours was. It is late in the game for Joan and me, but still early enough for many readers of this book! <u>My final advice is to go! Go often with kids if you have them and with borrowed funds if you need them – but go!</u>

CHAPTER 25
THE PLEASURES OF FOREIGN TRAVEL AS A SCIENTIST

I have traveled often enough to foreign countries to recognize one fundamental and significant difference between travels for scientific purposes and those for ordinary tourist objectives. Traveling scientists often if not usually look to counterparts, peers, associates or professional friends, for all kinds of assistance (which is usually given freely) whereas ordinary tourists must usually look to paid service people for sources of information, comfort, and well-being! This is a much larger advantage for the scientist than might be considered, and should be exploited fully in planning any kind of foreign travel – both for its scientific benefit as well as its comfort value! Visiting scientists can almost always expect to find a welcoming helpful scientist or group of scientists in foreign locations who are willing to provide advice and assistance.

Knowing that this personal relationship can exist serves to make travel for scientific purposes so much more meaningful than the usual tourist experience in the hands of paid service representatives. Visiting scientists expect to be welcomed and assisted by their foreign counterparts, as part of a mutual assistance program – formal or informal – but always reciprocal. The very same expectations exist among foreign scientists visiting the United States. They fully expect, and have every right to expect, the utmost in reciprocity when they arrive here. What immediately comes to mind in this instance was a somewhat unexpected visit by a South Korean Laboratory Director and his wife during our tenure in Miami. They placed themselves, without notice, in our hands for a weekend before their return home after a visit in several northern states. After some scrambling around we arranged a tour of the city and the laboratory, and an evening's entertainment and lodging at a Key Biscayne hotel – a typical South Florida experience, for which they seemed grateful. Examples of mutual assistance with travel issues are probably best found among members of international advisory groups – scientific working groups or committees – formed to assist with technical affairs at statutory meetings of international treaty organizations that meet regularly.

Scientists play significant but often largely unrecognized roles as participants in technical study and discussion groups or committees that report to or otherwise provide scientific information to, the designated official members of diverse international governmental regulatory bodies, formed by treaties or other agreements among nations or through United Nations Commissions.

Much of the development and provision of such technical information is the responsibility of government scientists and government laboratories, as well as of academic scientists encouraged through government grants to perform research activities that resolve specific problems – for example setting catch limits on whales, seals, tuna, and salmon populations.

Good scientific data are required as foundations for sustainable resource extraction. Acquisition of such data is expensive and must be innovative for best management of resource populations under stress from human activities. Population responses to human predation and other negative influences must be assessed, with the best scientific methodology available – and this is often <u>international</u> in scope.

Directors of laboratories whose research programs are planned to provide just such useful scientific data are frequently deeply involved in foreign meetings of the kind just considered – meetings that may take place in laboratories or cities of other member countries, and which in the long-term may create cadres of international scientists.

The results of the combined research efforts of several nations are frequently necessary to provide best technical advice by advisory groups often composed of the most knowledgeable senior scientists and laboratory directors from those nations – meeting, discussing, arguing, and usually finally agreeing on a course of action or at least a recommendation for the parent non-technical council or commission to consider.

I have been a member of several such "cadres" during my scientific career. Among them were those associated with the International Council for the Exploration of the Sea (ICES), the UN (FAO) International Commission for the Northwest Atlantic Fisheries (ICNAF), the U.S. Japan Joint Panels on Natural Resources (UJNR), and the UN (FAO) Commission on the Exploitation of Central African Fisheries (CECAF). All of these interactions have been intense learning experiences on large world stages, for which I am forever grateful. That kind of exposure to international science has been a priceless career-long experience for me!

But the pleasures of international science are certainly not limited to group activities associated with providing scientific advice to governmental or other statutory meetings. Scientists also form <u>disciplinary and multi-disciplinary communities</u> that can be national or international in their memberships, and can be particularly supportive for individual scientists venturing into foreign countries, possibly for the first time, but later as a matter of standard practice.

With networks throughout the world based on single scientific disciplines or aggregates of closely related disciplines of science, there is no need for any scientist to travel abroad as a simple tourist (unless that is his or her unlikely preference). There should always be a knowledgeable colleague, almost anywhere in the world, to smooth the path and offer advice to the scientist/traveler.

To conclude this consideration of friendships as an important ingredient of science, I would make the observation that many of my most meaningful and long-lasting friendships have been with other scientists, and frequently with foreign scientists. This may well be the outcome of a commonality of outlook and purpose in life, or a partial consequence of training and attitudes, but it is a very real phenomenon in my life, and I welcome it!

Readers of this book have discovered early on that Joan and I had five children early in our marriage and careers. Those kids went to school and college while we were busy with careers and bills and travel, but we did not take them on business-related foreign trips. We traveled with the family by car on visits and vacations, but we never took them on foreign trips – possibly because of conflicts with school schedules or lack of funds, or both. We did take summer vacations, but not outside the United States.

I bring up this serious omission so late in the narrative to be certain that it is apparent to every reader. The resulting message that I offer is, "<u>Always consider taking your children over 7 years old on any foreign trips that you make</u>," regardless of financial or other limitations, and regardless of the purpose of the trip! Borrow heavily if you must, but do it!

CHAPTER 26
THE BEST OF ALL TRAVEL DESTINATIONS:
THE UNITED STATES

All the previous discussions of foreign places of interest should not obscure the many locations within the United States that have been, for us, wonderful to visit officially or unofficially or both. As natives of an inland tourist area – the Berkshire Hills of western Massachusetts – Joan and I were familiar with tourist ways, but we still enjoyed their repetition with many nuances throughout the coastlines and interior of this great country of ours.

Since my research interests have always been coastal and ocean oriented, it was logical that the almost endless coastlines of the United States would receive our greatest attention, from Maine southward to Florida and the Gulf of Mexico, and then up the Pacific Coast to Alaska. Greater emphasis was always on the Atlantic coast, where my duty stations were, but the years were punctuated with frequent visits to other laboratories and locations on all coasts for meetings and conferences (Hawaii and Alaska were also included in our chain of federal fisheries research laboratories). Stopovers on trips to technical meetings made many inland areas of the United States available as well.

To most of us who are scientists, the meeting itself, regardless of its nature, provides the incentive and the motivation to attend or not to attend, regardless of its location. To many foreign scientists though, meetings in the United States are prizes, to be considered of high value. This seems to be in part because of desire for exposure to U.S. science in concentrated form, but also because of opportunities to visit at least a few of the many famous or unusual sites in this country before or after the meetings – places where world events have occurred – or places where famous world figures have held the stage.

We, of course, do the same things when we visit their countries, but it just seems that we here in this country have more places and more miles to cover than they do (which is true, of course, in most cases).

What this factoid should mean to us as United States scientists is that <u>we should take every advantage during periods of travel for scientific purposes to explore the wonders of the country we live in</u>. I can think of very few foreign places, for example, where coastal laboratories exist and technical meetings are held, that do not have locations or events of compelling interest nearby – certainly enough to warrant a day's (or more) investigation before or after the official meeting. I know that such an attitude toward exploration brought large satisfactions to us (always on our own expense account of course). I do feel, though, that for most scientists, some limited personal exploratory opportunities should be incorporated in plans for any scientific travel ventures within the United States. This country is too big, too varied, and too exceptional to do otherwise!

Figure 19. The San Francisco cable car at the bayside end of its run.

Too many scientists travel great distances within this country and barely see or experience its wonders at all! Truly exceptional meeting locations become merely "meeting places" or "conference sites" to be entered by plane and cab and left by cab and plane, with no thought or time allotted to the unique beauty or the historic significance of the surroundings. (I know this is true because I have been guilty of the same awful professional travel practices myself, to my lasting chagrin, and I have watched countless other scientists following the same wasteful hollow paths through what should have been exciting and enlightening journeys.)

This practice should be stopped! My solution is this: "Every scientific meeting should be <u>summarized</u> and <u>evaluated on the scene</u> the day <u>after</u> the meeting ends, to ensure that the funding institution receives full value for paying for the travel. This evaluation on scene should be accomplished (in a half hour or more) the morning following the close of the meeting and a <u>mandatory free day for sightseeing</u> should be provided in travel funding for that day. This will force the most focused of scientists to at least take a local tour rather than rushing for his or her plane. He or she may even decide to stay an extra few days (on their own time of course) and it (the free time in a meeting city) could become habit forming!

The concluding thought and hope here is to travel for scientific purposes, when it seems useful or advantageous to do so, and whether it be domestic or foreign, but <u>do so with some broader perspective on the environment that you will encounter</u> – the people and places, domestic or foreign. Opportunities for travel, domestic or foreign, are favored elements of many professional positions, as well as being subjects of private initiatives. Go!

P.S. Have you seen Yellowstone Park, or the Grand Canyon, or Niagara Falls? Have you climbed Mount Katahdin in Maine or Mount Washington in New Hampshire? Have you toured official Washington, D.C. by night? Have you flown by helicopter at night over New York City? Have you hiked a part of the Appalachian Trail in New England in Autumn? Have you seen an entire field of sunflowers or tulips in bloom? Have you ever taken a vacation in Hawaii or Florida? Opportunities for travel, domestic or foreign, are potential benefits of many professional positions. Go!

CHAPTER 27

LEADERSHIP ROLES FOR SCIENTIFIC RESEARCH LABORATORY DIRECTORS IN INTERNATIONAL SCIENCE

A scientific research laboratory director's role is a complex one both internally, within the laboratory walls, and externally, on local, regional, national and international levels of responsibility.

This book attempts to describe one research laboratory director's personal journey to various scientifically significant locations in North America and around the world – a world connected, as it turns out, by threads of scientific research emerging from fixed centers of intensive investigations called "scientific research laboratories," staffed with professionals equipped with advanced training and advanced degrees in relevant specialty areas of science. Of crucial importance to the well-being and productivity of this research entity is its head, usually identified as the laboratory director. This is normally a person, male or female, who is an excellent scientist, an effective administrator, and a good communicator to the larger scientific community and to the world.

In the course of a long career as a laboratory and research center director in ocean sciences, I have met with and enjoyed long term contacts with outstanding laboratory directors from many countries, and I have often been impressed with their qualifications for the job – which include a blend of technical and administrative skills. I have often been impressed also with the singular beauty of the environments in which their facilities were located and the symmetry of the research structures that they have created.

In an earlier book (Sindermann, 2012) I have described some exceptional directors; in this one I have emphasized the environments in which some of these directors perform their functions, and in which they develop an atmosphere conducive to excellent research. Together, the two books represent my best effort to describe the successful scientific research laboratory director and the research environment that he or she creates within the larger locale of the laboratory in a number of countries worldwide.

To any casual reader, the title of this book – suggesting serious travel involvement by a scientific research laboratory director – poses the immediate question of "Why?" "Why should a research laboratory director be anywhere officially except in his or her laboratory performing the everyday functions of "directing"? This entire book is an attempted response to that question, by considering in detail the true perimeters and responsibilities of the scientific research laboratory director's position – responsibilities that are worldwide, wherever related science is practiced on a substantial scale!

A scientific research laboratory director may be away from his or her office and occasionally in a foreign country for a number of good reasons, including (but not limited to) the following:

- To attend an international meeting of a professional society of which he or she is an active member, as a meeting organizer, session chairperson, presenter of a review paper, society officer, committee member, or member of the board of directors;

- To participate in annual research committee, or council, or working group meetings that are advisory to the many statutory intergovernmental groups that emerge from various international treaties and other forms of mutual agreements;

- To participate in planning and implementing an international symposium in the specialized area of science in which he or she is a recognized and significant figure;

- To participate personally in a short-term or long-term joint international research project, or to attend steering committee meetings of such projects;

- To attend the commemorative symposium or technical meeting designed to mark or celebrate the honoring, retirement, or death of a distinguished foreign friend who is or was a scientist;

- To receive an honorific medal or prize for technical or administrative achievement with international implications;

- To represent his or her laboratory at formal meetings of international non-profit organizations that may be contributing financially to the laboratory;

- To listen very carefully to other countries' progress reports in specific areas of science – especially research findings in specialties in which the director is also an expert;

- To visit, after appropriate invitations, several foreign laboratories that are achieving or have achieved favorable credits for specialized research related to that being done at the director's home laboratory in the United States; or

- To organize long-term agreements to conduct alternate year working visits to a foreign country for key members of the staffs of the home and the foreign counterpart laboratories. The visits could be in alternating years and locations, but they must engage a challenging research problem of genuine significance to the members of both national groups. Funding sources for such joint projects are of course critical, and must be clearly defined well in advance of any implementation. It must be clear from the outset that the research problem should be of approximately equal significance to both of the participating laboratories (for example epizootic disease-caused mortalities and effects on adjacent populations of the same fish species); and

- Undoubtedly the most pressing reason that scientific research laboratory directors spend time in foreign countries is that <u>those directors provide on the scene and highly plausible advice and leadership to the bureaucrats and politicians from the United States that actually participate in statutory meetings</u> of many kinds in many countries on any day of the year except Christmas. <u>The laboratory directors provide the correct level of scientific stature to be acceptable to their foreign counterparts (and hence to their official principals)</u> from other countries and ethnicities – any of whom may question the veracity of any statement that may be made for the record during the formal sessions!

So here, then, are some of my general thoughts about the extensive foreign and domestic travel that I have pursued and enjoyed as a scientific research laboratory director, and about some of the significant leadership roles that laboratory directors can play during meetings with foreign colleagues. The pleasures have always far outweighed the minor difficulties inherent in such travel, and I feel singularly privileged to have had the world as one of my immediate perspectives during my extended career in science! I have seen and appreciated countless occasions where scientific research laboratory directors have assumed leadership roles, and have performed successfully if not brilliantly, almost without exception, because of background, training, and experience. Many of them were Americans.

CHAPTER 28
ACHIEVING SUCCESSFUL INTERNATIONAL
SCIENTIFIC RELATIONSHIPS

A large part of foreign travel consists of visual and cognitive experiences, seeing and enjoying places of beauty and history. However, for most of us who are scientists, a significant part of the foreign experience involves interacting with professional people at all levels in all countries that are visited. This aspect of professional contact is especially important to the scientific traveler, whose purposes for being there often include satisfaction of some general professional objectives such as a meeting of a professional society or an advisory committee. Here the traveler should be able to relate closely and often for extended periods of time with foreign associates as well as others in supporting roles.

Much of the success and enjoyment of foreign travel, whether for scientific purposes or not, therefore, <u>depends on relationships with other people, with an added dimension for scientists</u>. This chapter concentrates on some of the human elements involved when scientists from the United States interact with those from foreign countries, either in their countries or at meetings here in the United States.

International scientific meetings may come in several varieties:

- An annual (or less frequent) meeting of an international professional society such as the World Aquaculture Society or the International Society of Parasitologists;

- An annual meeting of a working group or standing committee that reports to one of the members of statutory bodies like the International Council for Exploration of the Sea created by international treaties;

- An annual or periodic meeting of one of the many Commissions and working groups of the United Nations subsidiary bodies of the Food and Agriculture Organization (FAO);

- A one-time special international symposium devoted to a subject of world-wide interest (ex. Water, cholera, climate change, malaria, emergent viral diseases, etc.).

Regardless of the nature of the foreign scientific meeting, there are certain general guidelines that should govern behavior of scientists for each international event (whether held in the United States or in a foreign country):

A. <u>A new scientist who is invited to give his or her first paper at a foreign meeting of an international scientific society</u>. This selection implies that the scientist's work and publications have received attention beyond the laboratory, and that the invitee may be a candidate for future invitations, depending on performance, regardless of nationality. Science is of course a world-wide venture and good science can be international, so early and continuous participation in some of its broader geographic aspects can be rewarding and productive. That participation should begin early and can be fruitful.

B. <u>A mid-career scientist who has been invited to participate in an annual meeting in Europe of a working group or standing committee of a statutory international organization</u>. This invitation is or may be a somewhat belated recognition of the invitee's research and publications as being of possible utility on an international scale, possibly to contribute to solution of scientific problems of international dimensions. Nationality should not be a deciding factor, but I have found that much of the good science done in many foreign countries has been and is being done by members of such groups or at the request of those groups. My advice here is to do everything possible to maintain contact with such a group (or several of them) if your research permits it.

C. <u>A mid-career scientist who has been offered the chairmanship of a standing research committee of an International Scientific Society</u>. This offer of chairmanship is usually in recognition of many years of active participation in the governance and achievement of objectives of the society. Nationality should not be a factor.

D. <u>An offer by an international scientific society to a mid-career scientist to chair a special session in his or her specialty at the next annual meeting</u>. This offer should be considered as a signal honor for achievements in a scientific research field. The best and immediate approach to the job would be to form an international selection committee of specialists from foreign countries who could aid with the selection of participants, working by computer through various phases of selection of participants, but suggesting that the final selection would be yours.

E. <u>A senior scientist who has been offered the position of chairperson of a temporary international group or study group of a well-known worldwide scientific society;</u> He or she would be probably noted for his or her administrative as well as his or her professional competence, so scientific stature should not be a factor in selection, but he or she has undoubtedly long ago learned that <u>language and cultural differences exist</u> in such a group and they can never be ignored, even in a professional meeting room. Language and its proper transmission by interpreters is of principal initial concern, even though many of today's scientists have adequate comprehension of English and several other languages. This matter must be discussed frankly and at length by proposed members of a small scientific planning group with use of computers. The success of the working group is contingent in part on a satisfactory solution to the language issue!

Once the language problem has been confronted, the designated chairperson should immediately assemble a planning group to discuss terms of reference for the working/study group well before its first meeting (and assuming that selection of national members of that working group will be a prerogative of the countries participating).

So here then, to bring this chapter to some sort of conclusion, are some suggestions for achieving successful working relationships in international science. These working relationships are a primary determinant of the achievements and impacts of collaborative scientific research efforts on the future that we all share.

CHAPTER 29

GENERAL CONCLUSIONS ABOUT SCIENTIFIC TRAVEL IN FOREIGN COUNTRIES

So these are some of the places in the world that I (and in many cases my wife Joan) have visited and enjoyed. Our record is pretty good, but there are always other interesting areas left unseen. In my case, one such place was the South Polar region, which I missed by the narrowest of margins. I wanted to go, and the State Department had me on their provisional list of "Antarctic Scientific Observers," and I was even issued a parka and cold weather kit, but at the last minute I was replaced on the annual visiting committee by a scientist with a specialty in Antarctic birds – a much more relevant specialty to the Antarctica of the time than mine, which was epizootic diseases in fish and shellfish. The stronger specialty won, and correctly so. I didn't go! This whole interaction was part of an interesting international agreement to keep Antarctica free of commercial development by any country, but instead to conduct relevant research there, subject to international oversight.

But, coming back to where we <u>did</u> go, we liked most of the islands and island chains that we have visited, in some cases repeatedly, like the Hawaiian Islands, The Magdalen Islands, Helgoland, Prince Edward Island, and a host of others, but, always, it was the people and their attitudes that were determining factors. We also liked many parts of Western Europe, and fortunately, of all our travel time abroad, I would estimate that at least half of it was spent in Western European countries, largely through the good graces of the International Council for the Exploration of the Sea (ICES), with its many meetings.We also liked associating with scientific people in general, regardless of the location, probably because they were much like us in many ways except for place of birth. We have made good friends among scientific counterparts from other countries – especially among those we have seen repeatedly at advisory committee meetings that often meet annually (ICES for example).

Somewhere before the end of this fascinating tale of world travel – that of an inveterate traveler for legitimate and official scientific purposes – I must acknowledge the sacrifice of some of the transient and never to be recovered joys and sorrows of a growing family, during my solo travel to foreign countries. I was, for example, 900 hard road miles away, with no air transportation, on the Gaspé Peninsula when Joan (at home in Maine) began having premature birth pains and hemorrhages for birth of one of our children, and I didn't get back until the crisis was over! At a much later time, I was inextricably committed to chairing an important session of UJNR in Tokyo, when I received delayed notice that my three year old granddaughter Jenna had died of accidental poisoning 9000 miles away in Miami, Florida, and her funeral was that day. But, other than such major painful events that I have just described, for which there is no solace or excuse, or relief, there have been no other major painful aspects of the extensive travel described in this book as an acceptable part of a professional career, at least none that are apparent to me now. My concluding advice on this subject would be to <u>take as much of your family as often as possible on any travel -- domestic or foreign</u>!

(I would note further that I usually felt that my wife and children supported and were even excited about my international travel, even when I had to travel alone).

As a consequence of travel that has resulted in meeting and interacting with foreign scientists, my own perspectives on research and research administration have expanded significantly. I have learned, for example, that <u>informed scientific opinions, freely given and sincere, are rare and valuable gifts and not ones to be considered lightly</u>. I have also learned that the <u>opinions of scientific leaders (such as laboratory directors) are usually considered to be of importance by members of</u>

the general population, even if those opinions concern scientific subjects outside the director's specialty.

I have learned that a significant part of the science done in this country (apart from that done by industrial laboratories) has been designed and carried out to satisfy the needs of international obligations resulting from various treaties and agreements, or to satisfy the guidelines of private research foundations, or to satisfy the career plans of individual investigators.

I have also observed that, for scientists giving technical advice at statutory meetings of international organizations, the individual's stature and reputation as a scientist, as well as his or her position in the scientific hierarchy, are very important criteria for acceptance of statements made.

Probably the most important thing that I have learned from participating as a technical advisor in various statutory meetings is that the science involved is not always or even often a determining factor, but it must be present along with various political, social, or economic elements that may be of greater significance.

My recommendation to every scientist with a specialty that would be of interest to an international organization with matching interests would be to contact them with an expression of interest in their programs and a proposal for participation – with your curriculum and references.

My further advice for entry into foreign science would be to attempt a dual route: (1.) develop an association or joint research with a foreign individual counterpart, and (2.) find out about membership in working groups or study groups of international organizations in your specialty (maybe ask a present member to recommend you). Be knowledgeable, likeable, aggressive and persistent!

And so it turns out, so late in the day, that this book isn't about travel per se at all; it is rather a book about people and places that Joan and I have encountered during my scientific career outside the United States, with a few geographic facts and maps included as a framework for discussion.

I have already admitted to a career-long propensity for looking at science on a world-wide basis as well as an intense localized occupation. In that process I have stumbled on a number of insights, conclusions and perceptions that I would like to assemble in one final list here:

- Science is and should be international. Progress in science is shared and enhanced by an open literature and often shared research programs.

- Search for solutions to many planetary problems must be shared by cooperative international research programs.

- Much good science is accomplished in response to the needs of statutory bodies created by a host of international treaties and agreements, governmental or otherwise.

- Most of the major subdivisions, disciplines, and specialties of science have developed international societies as well as national societies. This trend should be encouraged by members and by governments.

- Scientific research laboratory directors are occupying, and should continue to occupy major roles in international science.

With this my story is at an end, and I leave the world and its wonders to future peripatetic marine scientists!

EPILOGUE

It is time now to conclude this account of some of our world travels, mostly for scientific purposes, as discussed throughout the book and brought to a focus in Chapter 27 on "Leadership Roles for Scientific Research Laboratory Directors in International Science."

I hope that my original plan of dividing our travels into five "Phases" based on my tenures at the base laboratories on the United States Atlantic coast provided some logical structure (along with the maps) for the presentation of geographic clusters of countries that we visited.

The traveling that we have done, and that I have tried to summarize in this book, has brought a realization of something that was never large in my mind – but should have been. The United States is truly a world leader in so many ways – apparent and less apparent. My experiences as scientific advisor to several international statutory organizations have brought out this conclusion clearly to me. I enjoyed foreign travel very much, but I especially enjoyed the meetings and conferences and actions that resulted from the United States' significant participation on a world stage – and a real down to earth one beyond newspaper headlines.

I have always assumed that science and scientists in the United States were as good as, or, in some cases better than, those in other developed countries of the world. My international travels, though relatively limited, have reinforced this belief, at least for the part of science that I feel competent to offer an opinion. But beyond this, it is the <u>cooperation</u> among scientists from different developed countries that was most impressive to me and that I appreciated the most.

I want to emphasize that I feel good about most of my official travel. Very often I have come away from foreign meetings with the good feeling that my presence there had made a difference, sometimes in a number of ways, but mostly positive ones. I felt that I was on top of my scientific specialty in most aspects, and that my inter-personal skills were improving with each encounter, although they were never perfect. Each foreign meeting was a learning experience for me, in many ways, although I seem to be slow at times to accept reality as others see it, and slow to change questionable personal habits as quickly as I should. Maybe, too, the relatively greater impact of a foreign professional trying to improve my style and attitudes could be a factor to be considered – and I did, on numerous occasions. At any rate, the entire foreign experience was a good one for me even though it seemed at the time to be part of a perfectly normal sequence of events! Beyond the joint professional activities however, was the genuine pleasure of association with knowledgeable and friendly people in so many of the countries we visited and revisited!

There are colleagues; especially in several Western European countries, whom we would particularly like to acknowledge for their efforts at welcome and friendship. They have made our travels doubly pleasant and left lasting memories.

Like most travelers, Joan and I had favorite countries – like Japan, France, and Ireland that we were able to visit several times – but we enjoyed being in many other countries, too, for various reasons. I especially liked the scientific and personal interactions with foreign colleagues and we, Joan and I, enjoyed the many other personal experiences that foreign travel brought, for many years. It was a good and important element in our lives!

Signed: Carl J. Sindermann
Joan P. Sindermann

A FINAL NOTE FROM THE AUTHOR

What kind of book is this? It is only weakly a travel book, even though it does describe a number of interesting places, and it is equipped with maps. It is not a scientific book, although it discussed occasional scientific events; It is, in total, a book of reminiscences by a former scientific research laboratory director about the travels that were a necessary part of his position. The travels he encountered took him and sometimes his wife Joan to interesting parts of the world, many of which are described here. The motivations for such travel were both professional and personal, but the result, as the book tries to indicate, was a life and career replete with pleasures and satisfactions, technical and personal!

Carl J. Sindermann

ACKNOWLEDGEMENTS

I want to acknowledge in particular, of the many people who helped me in my career in science, the six who were in my opinion, and for different reasons, most important.

1. Mrs. Joan Sindermann, wife and friend;

2. Dr. Gilbert L. Woodside, Chairman of the Department of Zoology, University of Massachusetts;

3. Mr. Leslie W. Scattergood, Director, Boothbay Harbor Fisheries Research Laboratory;

4. Mr. John B. Holston, Assistant Center Director, Middle Atlantic Fisheries Research Center;

5. Dr. Lionel A. Walford, Director, Sandy Hook Laboratory; and

6. Dr. Jay R. Traver, Assistant Professor, Department of Zoology, University of Massachusetts.

I thank each one of the above in particular:

1. Joan for lifelong enthusiastic support for every reasonable proposal that I presented to her (and some that were marginal).

2. Dr. G.L. Woodside for insisting that I apply for admission to graduate school at Harvard.

3. Mr. L.W. Scattergood, Director of the Boothbay Harbor Fisheries Research Laboratory, for tolerating the varied research initiatives of a new Harvard Ph.D.

4. Mr. John B. Holston, a wise and experienced scientist and Deputy Director of the Middle Atlantic Coastal Fisheries Research Center, for his effective leadership during my occasional absences from the Center;

5. Dr. Lionel A. Walford, my predecessor at Sandy Hook for presenting to me an excellent role model of an international scientist in every way; and

6. Dr. Jay R. Traver, for her extensive introduction to me of the methodology of scientific research, so long ago.

These are the people who were present and supportive at critical points in my career, and to whom I owe the utmost gratitude. This does not imply that there were not many others, but, in particular, I would mention Dr. John B. Pearce and Dr. Aaron Rosenfield, who were both important during phases of my career in science:

Dr. Aaron Rosenfield, a good friend since we team-taught a great general education course in Biology to an auditorium full of sophomores at Brandeis University more than half a century ago.

Dr. John B. Pearce, a good friend and a senior biologist at the Sandy Hook Laboratory, during my tenure there and subsequently at the Woods Hole Laboratory.

I thank you all for the many good things you have done for me!

(A SPECIAL ACKNOWLEDGEMENT TO MY FAMILY MEMBERS)

I want to acknowledge, in particular, the assistance of my family members for their roles in the preparation of this book. My daughters Nancy Sweet and Jeanne Kennedy provided detailed reviews and comments on every manuscript chapter; my son James Sindermann took responsibility for the cartography, and my sons Carl and Dana Sindermann took responsibility for the many complex logistic aspects of producing the entire book manuscript. My wife Joan of course provided many of the historical details that I have included. Sandra Rosenfield, daughter of my best friend, Dr. Aaron Rosenfield, typed the manuscript for publication. I thank you all, and I hope that the resulting book meets your expectations.

So, this view of the world has been a genuine family effort, and one that has given me much pleasure as well as insights in its preparation. What I have described is an aspect of a chosen way of life, and one that I have found more than satisfactory for one lifetime!

AUTHOR AND PUBLICATIONS

Carl James Sindermann

Among marine biologists and parasitologists around the world, the name Carl Sindermann is almost synonymous with books and articles on diseases of marine organisms. A background in fish and shellfish pathology, coupled with a talent for writing, propelled him into a successful career as a scientific author. Carl is a prolific and thoughtful writer, whose output includes not only scientific treatises, but books on the discipline of science and the scientists who practice it. Volumes that he has authored or co-authored are indispensable components of marine biology libraries worldwide.

Carl grew up in North Adams, Massachusetts, where he was born on August 28, 1922. After graduating from high school and working for two years for Pratt and Whitney, an airplane engine manufacturer in Hartford, Connecticut, he joined the U.S. Army at the outbreak of World War II. Carl served as a medic in an infantry reconnaissance platoon, landing in Normandy a month after D-Day and moving with Patton's army through France, Germany, Austria, and Czechoslovakia before the war ended.

In 1946, Carl enrolled at the University of Massachusetts on the GI Bill. It was not until his senior year, however, that his interest in science was galvanized. A female faculty member, who, Carl recalls, was still an assistant professor after 20 years in the Zoology Department, assigned him a research project that became his senior honor's thesis: the study of an invasion of western Massachusetts by a large, predatory land planarian. Carl worked out the life cycle of this flat worm, which had been imported in soil from the tropics and was destroying natural earthworms in greenhouses – and launched into a career as a parasitologist. He had already been accepted at Purdue University when the head of the Zoology Department at the University of Massachusetts took Carl to visit colleagues at

Harvard University. Carl was accepted on the spot to pursue a graduate program in parasitology.

Carl studied with the protozoologist, L.R. Cleveland, working on life cycles of parasites of wood-eating cockroaches. He shifted to the marine field for his Ph.D. research, working with mycologist W.H. Weston. Carl's Dissertation was based on summer research at the US Fish and Wildlife Service Laboratory at Boothbay Harbor, Maine, where he studied a fungus disease of herring. While at Harvard, Carl became a teaching assistant in Parasitology and Tropical Public Health at the Harvard Medical School, and also an instructor at nearby Brandeis University, where he taught undergraduate courses in biology and invertebrate zoology. He also continued his association with the U.S. Fish and Wildlife Service, serving as the Chief of the North Atlantic Herring Investigations. After obtaining his Ph.D. from Harvard in 1953, Carl remained on the Brandeis faculty until 1956, when he elected to return to the Boothbay Harbor Laboratory and become a research biologist.

Carl remained at Boothbay Harbor until 1962, by which time the Laboratory, along with all U.S. fisheries programs, had been transferred to the newly created Bureau of Commercial Fisheries (BCF). Carl's administrative skills were recognized within the Bureau, and in 1963 he moved to Maryland's Eastern Shore to become Director of the new laboratory at Oxford (now the Cooperative Oxford Biological Laboratory). The Oxford Laboratory was built as a consequence of the epizootic mortalities of eastern oysters, caused by MSX disease, that had begun in the Delaware and Chesapeake Bays a few years earlier. The new laboratory specialized in disease studies of commercially important fish and shellfish, and under Carl's direction, its scientists played important roles in the early days of oyster disease research and the laboratory's reputation became known world-wide. In 1968, Carl left the mid-Atlantic for a new post as director of the BCF's Tropical Atlantic Biological Laboratory in Miami, Florida, a job that he held for the next 4 years. In 1971, Carl returned to the mid-Atlantic, this time to New Jersey, where he became director of the Middle Atlantic Coastal Fisheries Research Center with headquarters at the Sandy Hook Laboratory of the National Marine Fisheries Service (NMFS – the old BCF). While serving in these posts, Carl had written not only numerous articles and reports, but had also become a renowned book author. In 1988, he withdrew from administration to devote full time to writing. From 1988 to 1990, he was an Intergovernmental Personnel Act Appointee, first at the University of Miami and later at the Maryland Department of Natural Resources, after which he returned to the Oxford Laboratory.

Throughout his career as a Federal employee, Carl retained close ties to academic institutions near his various postings. He held visiting or adjunct professorships at Georgetown, Florida Atlantic, Lehigh, and Cornell Universities, and the Universities of Miami, Guelph, and Rhode Island, where he taught invertebrate zoology, marine biology, fish pathology, and marine parasite ecology. He has served on the editorial boards of *Aquaculture, Chesapeake Science,* the *Journal of Fish Biology,* the *Journal of Invertebrate Pathology,* and the *Proceedings of the National Shellfisheries Association.* He was the Scientific Editor of the *Fishery Bulletin.*

The honors and awards that Carl has received are too numerous to list, but a sampling shows the breadth of activities and interests that have occupied him during the past half century: member, Bureau of Commercial Fisheries advisory group to NASA on back contamination from lunar exploration, 1967; recipient of the Department of Commerce Silver Medal for administrative and research activities, 1975; chairman, New Jersey Sea Grant Advisory Board, 1981-1985; keynote speaker for the Sixth Symposium on Pollution and Physiology of Marine Organisms, Charleston, SC, 1983. He served as the President of the World Mariculture Society in 1980-1981, and was chosen as an Honored Life Member of the National Shellfisheries Association 1991.

Although Carl has been a member of various international fisheries bodies, his work with the International Council for the Exploration of the Sea (ICES) is perhaps the most important. His affiliation with that organization began in 1959 when he attended his first meeting in Copenhagen. In the 1970s and 1980s he served on a number of ICES working groups, including those for Fisheries Improvement, Marine Aquaculture, Marine Pathology, and Introduced Species (of which he was chairman for a decade). His ability to synthesize and analyze great quantities of material was critical to the preparation of numerous reports for these groups, some of which served as the basis for later publications. An important contribution of these working groups was the issuance of the ICES "Code of Practice," which lists steps to be taken during the transfer of aquatic species to reduce the risks of disease spread when aquatic organisms are moved to new locations. The guidelines are used at present by most European countries and many US states.

Although Carl devoted much of his career to laboratory administration, he is best known as a book author. His scientific writing began as papers describing his research on marine parasites and pathology. His first publication (1953) was on "clam digger's itch," a human problem, but caused by a trematode parasite with a marine snail intermediate host. Carl's interests subsequently turned to parasites of the marine organisms themselves. Because he was in charge of the Atlantic herring project, his studies focused on this species, with a number of publications in the 1950s describing parasites and diseases of herring. Several of Carl's early papers showed that parasites could be used as tags to trace the movement of fish stocks. At the Boothbay Harbor Laboratory, Carl's work on serology of fishes resulted in a series of papers ranging from comparative serotyping of different fish species to the effects of disease on blood characteristics.

In the early 1960s Carl's ability to synthesize material became evident in an article entitled "Disease in marine populations" (1963). Not long afterward, he teamed up with Oxford Lab colleague Aaron Rosenfield, whom he had met while both were on the faculty of Brandeis University, to produce the now classic paper "Principal diseases of commercially important marine bivalve Mollusca and Crustacea," published in the Fishery Bulletin in 1967. In 1970, Carl expanded his earlier work in a volume entitled "Principal Diseases of Marine Fish and Shellfish" (Academic Press), which won the Wildlife Society of America award for best scientific publication in fisheries for 1970. Carl later updated this important work, which was re-issued in a 1990 two-volume set. These publications are acclaimed not only for the breadth of material included and the depth of analysis, but for the clarity of language and illustrations. His growing interest in aquaculture led to another indispensable book for the aquatic pathologist: "Disease Diagnosis and Control in North American Marine Aquaculture," edited by Carl and published in 1977 by Elsevier. This volume was also updated, in 1988, and in collaboration with Dr. Don Lightner.

While director of the Sandy Hook Laboratory, which is situated on the shore of the New York Bight, it was natural, perhaps inevitable, that Carl's attention would be drawn to the effects of coastal pollution on marine organisms. Once again, he meshed his sure-handed grasp of disease processes in the marine environment with what he was learning about pollution in a series of publications showing links between environmental contaminants and disease in marine fish. One of his most recent books, "Ocean Pollution Effects on Living Resources and Humans" (1996, CRC Press) is an outgrowth of these concerns.

Carl's enthusiasm for writing has led him into areas not often entered by scientists: writing about the scientists themselves. His first foray, titled "Winning the Games Scientists Play" (1982, Plenum) elicited enthusiastic reviews, and some consternation among a few colleagues who recognized themselves in the vignettes he used as illustrations. "The Joy of Science" (Plenum) was published in 1985, followed in 1987 by "Survival Strategies for New Scientists" and in 1992 by "The Woman Scientist" (co-authored by Clarice Yentsch). Carl's more recent offering, written in collaboration with Tom Sawyer, is entitled "The Scientist as Consultant:

Building New Career Opportunities" (1997). The books show Carl to be a keen observer of scientists and an accurate reporter of their behavior. They have an underlying theme: to analyze, often with a lighthearted touch, what makes a person successful in the scientific profession. They are realistic, discussing both pros and cons of certain career paths, and contain a wealth of practical advice – valuable not only for those considering or just embarking on a scientific career, but with admonitions that individuals well along in their professions would do well to follow.

Carl and his wife Joan are the parents of two daughters (both social scientists) and three sons (all with careers of their own). When Carl retired in 1990, he and Joan decided to remain on the Eastern Shore of Maryland – a decision that they have never regretted.

Since his retirement from the Federal Government in 1990, Dr. Sindermann has published several books: a revision of his 1982 classic "Winning the Games Scientists Play" in 2002, "Coastal Pollution" in 2006, and "The Scientific Research Laboratory Director" in 2012. His new book offering – "World Travels with a Peripatetic Marine Scientist" should be published in 2016.

PUBLICATIONS

SINDERMANN, CARL J.: Books Published (Author or Editor)

Sindermann, C.J. (in press). World Travels with a Peripatetic Scientific Research Laboratory Director.

Sindermann, C.J. 2012. The Scientific Research Laboratory Director. Xlibris Corp., Bloomington, Indiana. 203 pp.

Sindermann, C.J. 2006. Coastal Pollution: Effects on Living Resources and Humans. CRC Press, Boca Raton, FL., 280 pp.

Sindermann, C.J. and T.K. Sawyer 1997. Scientists as Consultants. Plenum, New York, 341 pp.

Sindermann, C.J. 1996. Ocean Pollution: Effects on Living Resources and Humans. CRC Press, Boca Raton, Florida, 275 pp.

Yentsch, C., and C.J. Sindermann. 1992. The Woman Scientist. Plenum, New York, 350 pp.

Sindermann, C.J. (Ed.) 1992. Special Symposium on Introductions and Transfers of Marine Species. ICES Mar. Sci. Series, Copenhagen, 174 pp.

Sindermann, C.J. 1990. Principal Diseases of Marine Fish and Shellfish (second edition). Acad. Press, New York, Vol. I, Fish, 536 pp., Vol. II, Shellfish, 513 pp.

Sindermann, C.J. and D.V. Lightner (Eds.). 1988. Disease Diagnosis and Control in North American Marine Aquaculture – second edition. Elsevier Scientific pubs., 431 pp.

Sindermann, C.J. 1987. Survival Strategies for New Scientists. Plenum, New York, 264 pp.

Sindermann, C.J. Proceedings of the Twelfth U.S. Japan Joint Natural Resources Panels, Reproduction, Maturation and Seed Production of Cultured Species. NOAA Tech Report NMFS 47, 73 pp.

Bilio, M., H. Rosenthal and C. Sindermann (Eds.) 1986. Realism in Aquaculture: Achievements, Constraints, Perspectives. European Aquaculture Soc., Bredene, Belgium, 585 pp.

Sindermann, C.J. 1985. The Joy of Science. Plenum, New York, 259 pp.

Sindermann, C.J. (Ed.) 1984. Proceedings of the Seventh US-Japan Joint Natural Resources Panel, Marine Finfish Culture. NOAA Tech Report, NMFS 10, 69 pp.

Sindermann, C.J. 1982. Winning the Game Scientists Play. Plenum, New York, 290 pp.

Sindermann, C.J. (Ed.) 1982. Proceedings of the Eleventh US-Japan Joint Natural Resources Panels, Salmon Enhancement. NOAA Tech Report NMFS 27, 102 pp.

Sindermann, C.J. (Ed.) 1980. Proceedings of the Ninth and Tenth US-Japan Joint Natural Resources Panels, Nutrition and Physiology. NOAA Tech Report NMFS 16, 92 pp.

Sindermann, C.J. 1979. Status of Northwest Atlantic Herrings Stocks of Concern to the United States. National Marine Fisheries Service, Sandy Hook Laboratory, Tech. Ser. Rept. No. 23, 447 pp. (Reprinted 1985, Maine Dept. Marine Resources) NTIS No. PB 80-189355.

Swanson, L. and C. Sindermann. (Eds.). 1979. Oxygen Depletion in the New York Bight – 1976. NOAA Prof. Pap. 11, 345 pp.

Kaul, P.N. and C.J. Sindermann (Eds.). 1978. Drugs and Food from the Sea. Univ. Oklahoma Press, Norman, 448 pp.

Sindermann, C.J. (Ed.). 1977. Disease Diagnosis and Control in North American Marine Aquaculture. Elsevier Sci. Pub. Co., Amsterdam, 329 pp.

Sindermann, C.J. (Ed.). 1974. Diagnosis and Control of Mariculture Disease in the United States. Middle Atlantic Coastal Fisheries Center, Technical Ser. Pub. No. 2, 306 pp.

Sindermann, C.J. 1970. Principal Diseases of Marine Fish and Shellfish. Academic Press, New York, 369 pp.

Sindermann, C.J. 1970. Bibliography of diseases and parasites of marine fish and shellfish. Tropical Atlantic Biological Laboratory, Miami, Florida, TABL Professional Report No. 11, 440 pp.

Sindermann, C.J. 1966. Diseases of marine fishes. T.F.H. Publications, Jersey City, New Jersey, 89 pp.

Journal Editor:

Proceedings of the US-Japan Joint Panels on Natural Resources (UJNR)

Fisheries Bulletin (U.S. National Marine Fisheries Service, National Oceanic and Atmospheric Administration (NOAA)

Printed in the United States
By Bookmasters